Lecture Notes in Physics

New Series m: Monographs

Springer-Verlag Berlin Heidelberg GmbH

The Editorial Policy for Monographs

The series Lecture Notes in Physics reports new developments in physical research and teaching - quickly, informally, and at a high level. The type of material considered for publication in the New Series m includes monographs presenting original research or new angles in a classical field. The timeliness of a manuscript is more important than its form, which may be preliminary or tentative. Manuscripts should be reasonably self-contained. They will often present not only results of the author(s) but also related work by other people and will provide sufficient motivation, examples, and applications.

The manuscripts or a detailed description thereof should be submitted either to one of the series editors or to the managing editor. The proposal is then carefully refereed. A final decision concerning publication can often only be made on the basis of the complete manuscript, but otherwise the editors will try to make a preliminary decision as definite as they can on the basis of the available information.

Manuscripts should be no less than 100 and preferably no more than 400 pages in length. Final manuscripts should preferably be in English, or possibly in French or German. They should include a table of contents and an informative introduction accessible also to readers not particularly familiar with the topic treated. Authors are free to use the material in other publications. However, if extensive use is made elsewhere, the publisher should be informed. Authors receive jointly 50 complimentary copies of their book. They are entitled to purchase further copies of their book at a reduced rate. As a rule no reprints of individual contributions can be supplied. No royalty is paid on Lecture Notes in Physics volumes. Commitment to publish is made by letter of interest rather than by signing a formal contract. Springer-Verlag secures the copyright for each volume.

The Production Process

The books are hardbound, and quality paper appropriate to the needs of the author(s) is used. Publication time is about ten weeks. More than twenty years of experience guarantee authors the best possible service. To reach the goal of rapid publication at a low price the technique of photographic reproduction from a camera-ready manuscript was chosen. This process shifts the main responsibility for the technical quality considerably from the publisher to the author. We therefore urge all authors to observe very carefully our guidelines for the preparation of camera-ready manuscripts, which we will supply on request. This applies especially to the quality of figures and halftones submitted for publication. Figures should be submitted as originals or glossy prints, as very often Xerox copies are not suitable for reproduction. For the same reason, any writing within figures should not be smaller than 2.5 mm. It might be useful to look at some of the volumes already published or, especially if some atypical text is planned, to write to the Physics Editorial Department of Springer-Verlag direct. This avoids mistakes and time-consuming correspondence during the production period.

As a special service, we offer free of charge LATEX and TEX macro packages to format the text according to Springer-Verlag's quality requirements. We strongly recommend authors to make use of this offer, as the result will be a book of considerably improved technical quality.

Manuscripts not meeting the technical standard of the series will have to be returned for improvement.

For further information please contact Springer-Verlag, Physics Editorial Department II, Tiergartenstrasse 17, D-69121 Heidelberg, Germany.

Hans-Jürgen Borchers

Translation Group and Particle Representations in Quantum Field Theory

 Springer

Author

Hans-Jürgen Borchers
Institut für Theoretische Physik
Universität Göttingen
Bunsenstrasse 9
D-37073 Göttingen, Germany

Cataloging-in-Publication Data applied for

Die Deutsche Bibliothek - CIP-Einheitsaufnahme

Borchers, Hans-Jürgen:
Translation group and particle representations in quantum field
theory / Hans-Jürgen Borchers.

(Lecture notes in physics : N.s. M, Monographs ; 40)
ISBN 978-3-662-14078-9   ISBN 978-3-540-49954-1 (eBook)
DOI 10.1007/978-3-540-49954-1
NE: Lecture notes in physics / M

ISBN 978-3-662-14078-9

Originally published by Springer-Verlag Berlin Heidelberg New York in 1996
Softcover reprint of the hardcover 1st edition 1996

The use of general descriptive names, registered names, trademarks, etc. in this publication does not imply, even in the absence of a specific statement, that such names are
exempt from the relevant protective laws and regulatons and therefore free for general
use.

Typesetting: Camera-ready by the author
Cover design: Springer-Verlag Design & Production
SPIN: 10481177   55/3142-543210 - Printed on acid-free paper

Foreword

At the time I learned quantum field theory it was considered a folk theorem that it is easy to construct field theories fulfilling either the locality or the spectrum condition. The construction of an example for the latter case is particularly easy. Take for instance an irreducible representation of the Poincaré group with positive energy, and as an algebra of observables all compact operators in that representation space. This algebra of observables is even an asymptotically Abelian algebra. Since it has only a single representation – except for multiples of this one – it is hardly possible to replace locality in order to obtain a theory with a reasonable physical structure. This example shows that it is not sufficient to replace locality by asymptotic Abelian-ness. The construction of a theory fulfilling locality without a positive energy representation was first done by Doplicher, Regge, and Singer [DRS]. However, modern investigations on the locality ideal in the algebra of test functions, started by Alcantara and Yngvason [AY], seem to indicate that this is a general feature; this means that most of the algebras of observables fulfilling the locality condition will not have representations that also fulfil the spectrum condition. This discussion shows that quantum field theory becomes a subject of interest only if both conditions are satisfied at the same time.

Some of the protagonists of the theory of local algebras would like to give a purely algebraic formulation of quantum field theory. However, for an algebraic formulation of the spectrum condition it must be possible to define Fourier transformations on the translation group. Therefore it is necessary that the action of this group be sufficiently continuous. If this is the case then the spectrum condition can be formulated on the algebraic level, as done by S. Doplicher [Dop], and moreover there are necessary conditions for the existence of a faithful representation obeying the spectrum condition [Bch70a]. Unfortunately this continuity assumption for the translations contradicts other requirements that also seem to be natural. If we look at the algebra of observables, associated with a finite region, then there are strong arguments for supposing that these are local von Neumann algebras. The existence of projections and the continuity assumption on the algebraic level do not match.

It is my opinion that such a continuity assumption for the action of the translation group is very strong. Many examples of thermodynamic systems show that continuity remains only in the time direction. Therefore one should try to get along without using this assumption. It is the subject of this text to find out how far one can handle the situation without the continuity assumption.

The contents of these lecture notes are not the general theory of local observables but a special object of this theory. For the general theory one should consult the book "Local Quantum Physics" by Rudolf Haag [[Haa]]. Here we will treat the translation group together with the positive energy condition. The interplay between the locality condition in configuration space and the spectrum condition in momentum space leads to remarkable properties for the translations. The Lorentz group will play no role in these notes. This is partly due to the fact that Lorentz symmetry is broken in electrically charged sectors [Buch,FMS] and partly due to the circumstance that it cannot be handled with the methods needed here. A reasonable treatment of the Lorentz group needs the Tomita–Takesaki theory [[Ta]].

The main tools for deriving properties of the translations are functional analysis and the theory of several complex variables. What is needed for the purpose of these notes will be presented in Chaps. II and III. The first chapter deals with the assumptions of the theory of local observables and the construction of simple examples. Properties of the spectrum of the translations, which are only consequences of the positive energy condition, are derived in Chap. II. In the last chapter we deal with those properties of the spectrum of the translations, which are consequences of the properties both in configuration and in momentum space. In addition we also derive some consequences for the theory of local observables.

These notes are based on lectures I gave in Göttingen several years ago. Some of my students went through my notes while preparing their diploma theses and corrected some misprints. I thank all of them for their help. In particular I would like to thank R.N. Sen from the Ben Gurion University of the Negev for reading the text and St. Trebels for helping with the figures.

Göttingen,
March 1996

Hans - Jürgen Borchers

Contents

I. The Axioms of the Theory of Local Observables   1
   1. The Assumptions of the Theory   1
   2. Examples: Free Bose Fields   7
   3. Examples: Free Fermi Fields   14
   4. Notes and Remarks   24

II. Translations and Spectrum Condition   28
   1. Relations Between Support Properties of Functions and Analytic
      Properties of Their Fourier Transforms   29
   2. Symmetry Groups and Continuity   35
   3. Spaces of Momentum Transfer   37
   4. Spectrum Condition: The One–Dimensional Case   41
   5. Characterization of Positive Energy States   48
   6. The Spectrum in a Cone   51
   7. Notes and Remarks   57

III. The Opposite Edge of the Wedge Problem   61
   1. Holomorphic Functions of Several Complex Variables   62
   2. The Edge of the Wedge Theorem   65
   3. The Double Cone Theorem   68
   4. The Jost–Lehmann–Dyson Representation   71
   5. Some Consequences of the Jost–Lehmann–Dyson Representation   79
   6. Applications: A Hole in the Spectrum   83
   7. Notes and Remarks   86

IV. Locality Condition and the Spectrum of Translations   89
   1. Commutators and the Edge of the Wedge Problem   90
   2. Locality and Minimal Translations   92
   3. Locality and the Shape of the Spectrum   98
   4. Invariant States and the Cluster Property   104
   5. Additivity Properties of the Spectrum   110
   6. Absence of Classical Fields   115
   7. Notes and Remarks   117

Bibliography   120

List of Symbols   126

Chapter I

The Axioms of the Theory of Local Observables

This chapter is devoted to a description of the assumptions of the theory. Since it is believed that the theory of local observables is a theory which starts more or less from first principles, our discussion also tries to start from some general ideas. However, no attempt is made to enter into the theory of measurement, or into a discussion about the justification of the standard setting of quantum mechanics. The only purpose of the first section is to give some plausibility arguments as to why the axioms are chosen in the manner presented.

In sections two and three models are constructed which fulfil the assumptions of the theory, in order to show that these axioms do not describe the empty set. These examples describe the non-interacting theory, the so-called free fields. The second section deals with the free Bose field and the third with the free Fermi field. Examples of interacting fields do exist in two and three space-time dimensions. For the time being there exist no models in four dimensions describing interacting particles on a mathematically rigorous basis.

I.1 The assumptions of the theory

In every physical theory one is dealing with two kinds of objects. One kind consists of things which one wants to investigate. They are called the states, and they are either given by nature or they are prepared for the sake of an experiment. The second class of objects consists of the devices used for investigating the states. They are called observables. The observables used here are idealizations of the measuring instruments of the experimental physicist. For measuring a certain physical quantity, the experimentalist will use a family of different devices, each applicable in a certain energy range, but our observables are applicable in all circumstances. In most cases the outcome of an experiment is a position on one or several scales, i.e. real numbers and units describing the scales. In our idealization we fix the units once and for all so that we are left with real numbers. We may think of the units as being attached to the observables. They are needed for computing numbers to be tested by experiments.

Let $\mathcal{S}$ be the set of states and $\mathcal{O}$ be the set of observables. It is assumed that every observable $x \in \mathcal{O}$ can be used in order to analyze any state $\omega \in \mathcal{S}$. The outcome of the measurements defines a map:

$$\mathcal{M}: \quad \mathcal{S} \times \mathcal{O} \Longrightarrow \mathbb{R}.$$

The value of the map will be denoted by $\omega(x)$.

It is clear that the measurement should fulfil some obvious properties, namely two states will be called identical if they give the same result on every observable, i.e.:

$$\omega_1, \omega_2 \in \mathcal{S} \qquad \text{and}$$
$$\omega_1(x) = \omega_2(x) \quad \text{for every } x \in \mathcal{O}$$
$$\text{implies}: \quad \omega_1 = \omega_2.$$

Correspondingly two observables will be called identical, if they give the same result when applied to any state:

$$x_1, x_2 \in \mathcal{O} \qquad \text{and}$$
$$\omega(x_1) = \omega(x_2) \quad \text{for every } \omega \in \mathcal{S}$$
$$\text{implies}: \quad x_1 = x_2.$$

These two properties mean that $\mathcal{S}$ separates $\mathcal{O}$ and vice versa. This scheme is flexible enough to allow future discoveries. What we call a state nowadays might turn out to be an equivalence class of states at later times. But this is only possible after having discovered new observables and new states at the same time because states and observables must be mutually separating.

In order to get to a theory it is necessary to put more structure on the sets $\mathcal{S}$ and $\mathcal{O}$. Since von Neumann's description of the quantum mechanics [[Neu]] it is customary to assume that the observables are self–adjoint elements affiliated with a von Neumann algebra or with a C^*–algebra. Here we only will use bounded elements, corresponding to bounded functions of "real" physical observables. The physical states are the normal states of the algebra in the first case and all states if we deal with C^*–algebras. The mathematical expression "state" means a normalized positive linear functional on the algebra. There have been many attempts to justify this choice. Most of them are known by the name of quantum logic. This kind of investigation was initiated by Birkhoff and von Neumann [BN]. A survey of the early part of this theory can be found in the book of H. Reichenbach [[Rei]]. For more modern discussion of this subject see Varadarajan [[Var]] or the book of R. Haag [[Haa]]. These investigations try to give a justification of quantum mechanics of a finite number of degrees of freedom. In this situation one can assume the existence of logical atoms. Later on they will become the minimal projections of the von Neumann algebra to be constructed. Since different people choose different axioms for their setting this theory is not in its final state. Therfore, in my opinion, these investigations are not yet conclusive. Morover, the procedure of canonical quantization

$$[q_i, p_j] = i\delta_{i,j}, \quad [q_i, q_j] = [p_i, p_j] = 0$$

does not assign a unique meaning to functions of q and p. But fortunately these ambiguities do not affect the great success of quantum mechanics.

The situation becomes even less satisfactory if we look at quantum field theory, i.e. at systems with an infinite number of degrees of freedom. All the nice results obtained by quantum logic using the existence of logical atoms cannot be used here. There exist no atoms since every subsystem can be enlarged. This implies that a logical atom of a subsystem does not remain a logical atom for the enlarged system. If we start from a Lagrange function the construction of a field theory (if it is possible) seems to be more ambiguious than the constructions in standard quantum mechanics. In the latter case one can start from the Poisson bracket, while the meaning of an infinite dimensional Poisson space is unclear because of the lack of a definite topology. Despite all these problems I will take a pragmatic attitude. So far the algebraic approach has been very successful and therefore I will stay with it.

This means it is assumed that the observables are self-adjoint elements of a $*$–algebra containing the identity. States are then identified with the normalized positive linear functionals on this algebra , i.e. states which take value 1 on the identity. The set of normalized positive linear functionals on a $*$–algebra forms a convex set and it will be assumed that the physical states coincide with this set. If moreover for every $x \in \mathcal{O}$

$$\sup\{|\omega(x)|; \omega \in \mathcal{S}\}$$

exists, then this defines a norm on $\mathcal{O}$, so that $\mathcal{O}$ is the self-adjoint part of a normed $*$–algebra. Since every state on $*$–algebra gives rise to a representation on a Hilbert space, it follows with some additional assumptions that the normed algebra is a dense subalgebra of a C^*–algebra. From now on we shall assume that the observables $\mathcal{O}$ form the self-adjoint part of a C^*–algebra $\mathcal{A}$ and that the set of states coincides with the states of the algebra $\mathcal{A}$.

Every measurement is performed in a laboratory. Moreover, it is done by a device which has to be switched on in order to be prepared for a measurement, and, after the measurement is over, the device will be switched off. Therefore one should associate to every observable x a domain $O(x)$ in space-time, where $O(x)$ is the smallest domain in which x can be measured. Then it is clear that x can be measured in every larger domain. We also assume that, if x can be measured in O_1 and y in O_2, then $x + y, xy + yx$ and $i(xy - yx)$ can be measured in $O_1 \cup O_2$.

In the following the space-time will always be a Minkowski space of dimension $d \geq 2$, this means the space $\mathbb{R}^d$ furnished with the metric $g_{i,j}$ with $g_{0,0} = 1, g_{i,i} = -1$ for $i = 1, \ldots d - 1$, $g_{i,j} = 0, i \neq j$. For $a, b \in \mathbb{R}^d$ the scalar product is defined by $(a, b) = a^0 b^0 - \sum_{i=1}^{d-1} a^i b^i = a^0 b^0 - (\mathbf{a}, \mathbf{b})$ with $a = \{a^0, \mathbf{a}\}$. The (open) forward light–cone V^+ denotes the set $a \in \mathbb{R}^d$ with $a^2 > 0$ and $a^0 > 0$. A point $a \in \mathbb{R}^d$ is called spacelike if $a^2 < 0$. With this notation the first axiom is the following:

Axiom I.

To every bounded open region O in Minkowski space is associated a C^*–algebra $\mathcal{A}(O)$ which fulfils isotony, i.e.,

$$O_1 \subset O_2 \quad implies \quad \mathcal{A}(O_1) \subset \mathcal{A}(O_2).$$

The meaning of this assumption is, of course, that the self-adjoint elements in $\mathcal{A}(O)$ are exactly all observables which can be measured in O.

By this axiom, the family of C^*–algebras $\mathcal{A}(O)$ form an increasing net. Its C^*–inductive limit will be denoted by $\mathcal{A}$. If G is an arbitrary open domain in $\mathbb{R}^d$ then $\mathcal{A}(G)$ will be the smallest sub-C^*–algebra of $\mathcal{A}$ containing $\mathcal{A}(O)$ for every bounded $O \subset G$.

This theory should describe not only quantum mechanics but also satisfy the principles of special relativity. One of the concepts of relativity is that an action can be transmitted with a velocity at most that of light. This implies that an event in the domain O can influence other objects only if they are in the domain

$$O + \overline{V}^+.$$

If we associate to every observable A a domain $O(A)$ in space–time then this observable can disturb other objects only if their location enters $O + \overline{V}^+$. Therefore, two observables do not influence each other if their supports are spacelike separated. But by the principles of quantum mechanics this means that the operators representing these observables must commute with each other. From this discussion follows:

Axiom II.

If two bounded regions O_1 and O_2 are spacelike separated (this means $a \in O_1$ and $b \in O_2$ implies always $(a - b)^2 < 0$) then every operator in $\mathcal{A}(O_1)$ commutes with every operator in $\mathcal{A}(O_2)$.

Physics is based on the fact that one can repeat experiments at different places and at different times. Such a situation is called a symmetry. In terms of the previous discussion a symmetry is a pair of mappings (α, β) such that

$$\alpha : \mathcal{O} \to \mathcal{O} \quad \text{maps observables onto observables}$$

and

$$\beta : \mathcal{S} \to \mathcal{S} \quad \text{maps states onto states}$$

without changing the results of measurements. This means for $\omega \in \mathcal{S}$ and $x \in \mathcal{O}$ one requires

$$(\beta\omega)(\alpha x) = \omega(x).$$

If we replace the observables by the hermitean part of a C^*–algebra and if α is also an affine map which leaves the identity fixed, then it can be shown that α may be decomposed into two parts where one part is an automorphism of the algebra and the other is an antiautomorphism. Here we are interested in a group of symmetries where every group element is represented by an automorphism (the use of anti-automorphisms for continuous groups is burdened with some pathologies). As an example we consider the time symmetry then α_t or β_t has to be interpreted as time development. Because of the two possible choices we get either the Heisenberg or the Schrödinger picture of quantum mechanics.

The group we are interested in is generated by the translations and the Lorentz transformations. For this we use the name Poincaré group. Amusingly enough the homogeneous transformations are seldom used in the theory of local observables. Therefore, we state the third axiom in two parts

Axiom III.

The translation group of the Minkowski space $\mathbb{R}^d$ acts as a group of automorphisms on $\mathcal{A}$ in such a way that the equation

$$\alpha_a \mathcal{A}(O) = \mathcal{A}(O + a), \quad a \in \mathbb{R}^d$$

holds for every bounded open region $O \subset \mathbb{R}^d$.

Sometimes it is useful to have the whole Poincaré group, denoted by P, acting as symmetry group, in which case we formulate the axiom as follows:

Axiom III.P

The Poincaré group of $\mathbb{R}^d$ acts as group of automorphisms on $\mathcal{A}$ such that

$$\alpha_g \mathcal{A}(O) = \mathcal{A}(O_g)$$

holds for every g in the proper orthochronous Poincaré group.

Here g consists of a Lorentz transformation Λ and a translation $a \in \mathbb{R}^d$ for which $g = (\Lambda, a)$ with

$$(\Lambda, a)x = \Lambda x + a, \quad \forall x \in \mathbb{R}^d,$$

and O_g denotes the image of O by g.

Before formulating the next axiom, some notation is needed. If we have a C^*–algebra $\mathcal{A}$ and a group G acting by automorphisms α_g on $\mathcal{A}$ then we call the triple $\{\mathcal{A}, G, \alpha\}$ a C^*–dynamical (kinematical) system. A representation $\{\mathcal{H}, \pi, U\}$ will be called a covariant representation if π is a non–degenerate representation of $\mathcal{A}$ on a Hilbert space $\mathcal{H}$, and U a *continuous* unitary representation of G on $\mathcal{H}$ implementing the automorphisms α_g, i.e., fulfilling the equation

$$U(g)\pi(x)U^*(g) = \pi(\alpha_g x), \quad x \in \mathcal{A} \quad g \in G.$$

If now G is the abelian group $\mathbb{R}^d$ then one has an integral representation

$$U(a) = \int \exp\{\mathrm{i}(a, p)\} \, \mathrm{d}E(p)$$

where $E(p)$ is a projection-valued measure on the dual space of $\mathbb{R}^d$ which is again $\mathbb{R}^d$. The group representation $U(a)$ is said to fulfil the spectrum condition if the support of the measure $dE(p)$ is contained in the closed forward light–cone $\overline{V}^+$. A representation of $\{\mathcal{A}, \mathbb{R}^d, \alpha\}$ is called a representation with spectrum condition if the spectrum of U in $\{\mathcal{H}, \pi, U\}$ is contained in the closed forward light cone $\overline{V}^+$. Such a representation will be denoted by $\{\mathcal{H}, \pi, U, \overline{V}^+\}$.

We are convinced that we can learn the laws of physics by investigating finite–particle physics. Infinite–particle physics does not exist in a laboratory. This means we believe that no new laws of physics are needed when passing to infinite systems, as for instance systems exhibiting thermodynamical behaviour. But the discovery of all physics means in particular the discovery of all observables. From this we conclude that all finite–particle representations together should give rise to a faithful representation of the algebra of observables.

But what are finite–particle representations? This is clear when dealing with a theory containing only massive particles, namely these are representations fulfilling the spectrum condition. However, if we have photons or other massless particles in the theory then in the same representation space where we have a finite number of massive particles there might exist also clouds of massless particles containing an infinite number of particles but with finite energy. Therefore one should associate the notion of finite–particle representations to representations with positive energy. But since we wish to have a theory fulfilling the principles of special relativity the finite–particle representations will then be identified with representations fulfilling the spectrum condition. From this discussion follows:

Axiom IV.

The algebra $\mathcal{A}$ generated by all local observables admits a faithful representation $\{\mathcal{H}, \pi, U, \overline{V}^+\}$ fulfilling the spectrum condition.

A theory satisfying Axioms I to IV (without Axiom III.P) will be called a theory of local observables, and will be denoted by $\{\mathcal{A}(O), \mathcal{A}, \mathbb{R}^d, \alpha\}$.

I.2 Examples: Free Bose fields

In order to show that the axioms are not contradictory one has to construct some examples fulfilling the assumptions of the last section. The models in this section are based on the canonical commutation relations (C.C.R.). Therefore we start with a short discussion of the C.C.R. before constructing the examples. The C.C.R. appear in several different forms. Some of them are as follows:

Case I:

Let E be a real vector space furnished with a non-degenerate symplectic form $\sigma(x, y)$. This means σ is defined on $E \otimes E$ and has the properties

$$\sigma(x, y) = -\sigma(y, x) \quad \text{and} \quad \sigma(x, y) = 0 \quad \forall\, y \in E \quad \text{implies} \quad x = 0.$$

One is searching for an abstract $*$–algebra $\mathcal{A}_{E,\sigma}$ such that $\mathcal{A}_{E,\sigma}$ is generated by the elements $u(x)$ fulfilling

$$u(x) = u(-x)^* = u(-x)^{-1}, \quad \text{and} \quad u(0) = \mathbb{1},$$

and

$$u(x + y) = \exp\{i\sigma(x, y)\}u(x)u(y).$$

Case II:

Assume E_1 and E_2 are two real vector spaces and $B(x, y)$, $x \in E_1$, $y \in E_2$ is a nondegenerate bilinear form on $E_1 \times E_2$. One is looking for a $*$–algebra $\mathcal{A}_{E_1, E_2, B}$ generated by the elements $u(x)$, $x \in E_1$ and $v(y)$, $y \in E_2$ with the properties

$$
\begin{aligned}
u(x) &= u(-x)^* = u(-x)^{-1}, & x &\in E_1, \\
[u(x), u(x')] = 0, \quad u(x)u(x') &= u(x + x'); & x, x' &\in E_1, \\
v(y) &= v(-y)^* = v(-y)^{-1}, & y &\in E_2, \\
[v(y), v(y')] = 0, \quad v(y)v(y') &= v(y + y'), & y', y &\in E_2,
\end{aligned}
$$

and

$$u(x)v(y) = \exp\{iB(x,y)\}v(y)u(x).$$

Defining the space $E = E_1 \oplus E_2$ and the symplectic form σ on E by:

$$\sigma(x \oplus y, x' \oplus y') = -\frac{1}{2}\{B(x,y') - B(x',y)\},$$

and introducing elements $\tilde{u}$ by

$$\tilde{u}(x \oplus y) = u(x)v(y)e^{-\frac{1}{2}B(x,y)}$$

one obtains by simple calculation

$$\tilde{u}(x' \oplus y')\tilde{u}(x \oplus y) = \exp\{-i\sigma(x' \oplus y', x \oplus y)\}\tilde{u}\{(x' + x) \oplus (y' + y)\}.$$

But this shows that II is a special case of I.

Case III:

If the space E of the case I happens to be a real Hilbert space, and if the symplectic form σ is a continuous bilinear form then the Riesz representation theorem tells us that there is a linear operator β acting on E with:

$$\sigma(x,y) = (x,\beta y).$$

This operator β has the properties:

$$\beta^* = -\beta \quad \text{and} \quad \beta^2 \leq 0.$$

Case IV:

Finally let E be a complex Hilbert space with scalar product (x,y). Then

$$< x,y > = \operatorname{Re}(x,y)$$

defines a non-degenerate real scalar product and makes E into a real Hilbert space. In addition

$$\sigma(x,y) = \operatorname{Im}(x,y)$$

defines a symplectic form on E, so that IV is a special case of III. The corresponding algrbra will be denoted by $\mathcal{A}_{\mathcal{H},(.,.)}$. On the other hand having the information of III one is able to construct a complex Hilbert space provided one has $\beta^2 = -1$. For $\lambda, \mu \in \mathbb{R}$ identify the vector

$$(\lambda + i\mu)x \quad \text{with} \quad \lambda x + \mu\beta x$$

and define a new scalar product by the equation

$$(x,y) = < x,y > + i < x,\beta y >$$

The algebra $\mathcal{A}_{E,\sigma}$, described in case I, consists of elements of the form $\sum \lambda_i u(x_i)$, $\lambda \in \mathbb{C}, x_i \in E$ and on this algebra one can introduce a norm:

$$\left\| \sum \lambda_i u(x_i) \right\|_1 = \sum |\lambda_i|.$$

Denoting by $\mathcal{A}_{E,\sigma}^1$ the completion of $\mathcal{A}_{E,\sigma}$ in that norm, one defines $\widehat{\mathcal{A}}_{E,\sigma}$ as the enveloping C^*–algebra of $\mathcal{A}_{E,\sigma}^1$. This is not the only way of defining C^*–algebras associated to the canonical commutation relations, but this one is the minimal one containing all $u(x)$.

In applications to physics not every representation of $\widehat{\mathcal{A}}_{E,\sigma}$ is of interest. One is looking for those representations π for which the unitary representation $\pi(u(tx))$, $t \in \mathbb{R}, x \in E$ is continuous in t for every fixed $x \in E$.

For constructing a model of the theory of local observables, let $m > 0$. Then $\delta(p^2 - m^2)\Theta(p_0)$ defines a measure on the upper sheet of the hyperboloid $p^2 = m^2$. Let $\mathcal{H}_m$ be the Hilbert space

$$\mathcal{H}_m = \mathcal{L}^2\big(\delta(p^2 - m^2)\Theta(p_0)\big)$$

and let

$$\Delta_m^+(x) = \mathcal{F}\big(\delta(p^2 - m^2)\Theta(p_0)\big)$$

be the Fourier transformation of the above measure. Since the measure has its support in the light cone $\overline{V}^+$ it follows (for details see II.1) that $\Delta_m^+(x)$ is the boundary value of an analytic function holomorphic in the forward tube

$$T(V^+) = \{z \in \mathbb{C}^d; \operatorname{Im} z \in V^+\}.$$

Moreover, since the measure is real and invariant under Lorentz transformations one obtains

$$\overline{\Delta_m^+(-\bar{z})} = \Delta_m^+(z),$$

and by the Hall-Wightman theorem, that $\Delta_m^+(z)$ is invariant under complex Lorentz transformations. This implies in particular that $\Delta^+(z)$ has the following special form:

$$\Delta_m^+(z) = F(z^2),$$

where $F(w)$ is holomorphic in the cut plane $\mathbb{C} \setminus \mathbb{R}^+$. (The two different boundary values of $F(z^2)$ correspond to $x \in \overline{V}^+$ and $x \in -\overline{V}^+$). From the two relations of Δ_m^+ we obtain

$$\Delta_m^-(x) := \Delta_m^+(-x) = \Delta_m^+(x) \quad \text{for} \quad x^2 < 0,$$

and consequently

$$\operatorname{Im} \Delta_m^+(x) = 0 \quad \text{for} \quad x^2 < 0.$$

If $f, g \in S(\mathbb{R}^d)$ then they can be mapped by Fourier transformation into $\mathcal{H}_m$. One obtains as scalar product:

$$\int \overline{f(x)} \Delta_m^+(x-y) g(y) \, dx \, dy = \int \overline{\mathcal{F}^{-1}f(p)} \mathcal{F}^{-1}g(p) \delta(p^2 - m^2) \Theta(p_0) \, dp.$$

Due to the fact that the measure is supported only on the upper branch of the hyperboloid one sees that already the real elements in $S(\mathbb{R}^d)$, denoted by $S_r(\mathbb{R}^d)$, define a dense subset of $\mathcal{H}_m$ (by Fourier transformation). Next we define the local algebras. Since The algebra is described by the hyperboloid of mass m we write $\mathcal{A}_m$ instead of $\{\mathcal{A}; \mathcal{H}, \Delta_m^+\}$.

I.2.1 Definition:

On $S_r(\mathbb{R}^d)$ define a scalar product by

$$(f, g) = \int f(x) \Delta_m^+(x-y) g(y) \, dx \, dy$$

which makes $S_r(\mathbb{R}^d)$ into a pre-Hilbert space. Let $\mathcal{A}_m$ be the C^*–algebra as above and define $\mathcal{A}(O)$ to be the C^*–algebra generated by $\{u(f); \operatorname{supp} f \subset O\}$. From this one obtains:

I.2.2 Lemma:

Define on the algebra $\mathcal{A}_m$ an automorphism α_g, $\quad g = (\Lambda, a)$ by

$$\alpha_g u(f) = u(f_g), \quad \text{with} \quad f_g(x) = f(\Lambda^{-1}(x-a)).$$

Then the system $\{\mathcal{A}(O), \mathcal{A}_m, P, \alpha\}$ fulfills the axioms I, II, and III.P of section I.1.

Proof: That Axiom I is fulfilled is trivial. Next let $f, g \in S_r(\mathbb{R}^d)$ and $\operatorname{supp} f \subset O_1, \operatorname{supp} g \subset O_2$ and assume O_1 and O_2 are spacelike separated. Since f and g are both real we obtain

$$\operatorname{Im}(f, g) = \frac{1}{2i} \int f(x) \{\Delta_m^+(x-y) - \overline{\Delta_m^+(x-y)}\} g(y) \, dx \, dy = 0$$

since $\operatorname{Im} \Delta_m^+(x-y) = 0$ for $(x-y)^2 < 0$. From this one sees that $u(f)$ and $u(g)$ commute.

From the invariance of the measure under Lorentz transformations one finds that the equation $\Phi(f) = \Phi(g)$ implies $\Phi(f_h) = \Phi(g_h)$, with $h = (\Lambda, a)$, and where Φ denotes the mapping of $S_r(\mathbb{R}^d)$ into $\mathcal{H}_m$ defined above. This shows that the above map indeed defines an automorphism of $\mathcal{A}_m$ satisfying Axiom III.P. $\qquad\square$

In order to show that one obtains a local ring system one has to construct a representation of this algebra fulfilling the spectrum condition. This will where $< ., . >$ denotes the scalar product of the real Hilbert space.

be the Fock representation of $\mathcal{A}_m$. Denote by $S^n\mathcal{H}_m$ the symmetric part of the n-th power of $\mathcal{H}_m$ with respect to the tensor product. Define

$$S\mathcal{H}_m = \sum_0^\infty \oplus S^n\mathcal{H}_m, \quad S^0\mathcal{H}_m = \mathbb{C}.$$

Since the construction presented here is independent of the Hilbert space, $\mathcal{H}$ might be any Hilbert space without further specification until the representation of the Poincaré group is needed.

I.2.3 Lemma:

For every $\psi \in \mathcal{H}$ define in $S\mathcal{H}$ a special vector which will be denoted by $\exp\psi$.

$$\exp\psi := 1 \oplus \psi \oplus \frac{1}{\sqrt{2!}}\psi \otimes \psi \oplus \cdots \oplus \frac{1}{\sqrt{n!}}\psi^{\otimes n} \oplus \cdots \quad .$$

These vectors have the properties:

$$(\exp\psi, \exp\varphi) = e^{(\psi,\varphi)}$$

and, moreover, the set of vectors $\{\exp\psi, \psi \in \mathcal{H}\}$ is total in $S\mathcal{H}$.

Proof: The relation for the scalar products is a simple consequence of the definition of tensor products of Hilbert spaces. To prove the second statement take $\varphi \in S\mathcal{H}$ with $(\varphi, \exp\psi) = 0$ for all $\psi \in \mathcal{H}$. Writing $\varphi = \sum \varphi_n$ with $\varphi_n \in S^n\mathcal{H}$ one has $\sum \|\varphi_n\|^2 < \infty$. For $z \in \mathbb{C}$ one finds:

$$|(\varphi, \exp z\psi)| \leq \sum |(\varphi_n, \psi^{\otimes n})\frac{z^n}{\sqrt{n!}}| \leq$$

$$\sum \|\varphi_n\| \|\psi\|^n \frac{|z|^n}{\sqrt{n!}} \leq \{\sum \|\varphi_n\|^2\}^{1/2} e^{1/2\|\psi\|^2|z|^2}.$$

This implies $(\varphi, \exp z\psi)$ is an entire analytic function in z and consequently $(\varphi, \exp z\psi) = 0$ implies $(\varphi_n, \psi^{\otimes n}) = 0$. Inserting for ψ the vector $\sum_{i=1}^n w_i\xi_i$ with $\xi_i \in \mathcal{H}$ and $w_i \in \mathbb{C}$ one obtains a polynomial of degree n in w_i. Picking the coefficient of $\prod_{i=1}^n w_i$ one obtains

$$(\varphi_n, S^n\xi_1 \otimes \xi_2 \otimes \cdots \otimes \xi_n) = 0$$

where S^n denotes the symmetrization operator. But these vectors are total in $S^n\mathcal{H}$ so $\varphi_n = 0$ and hence $\varphi = 0$. □

Using this notation one defines a representation of the Weyl algebra as follows:

I.2.4 Definition:

Let $\mathcal{H}$ be furnished with the symplectic form $\sigma(x,y) = \mathrm{Im}\,(x,y)$; on $S\mathcal{H}$ one defines linear operators $\pi(u(x)) =: U(x)$ by the equation:

$$U(y)\exp x = e^{\{-1/2\|y\|^2-(y,x)\}} \exp(y+x).$$

I.2.5 Lemma:

Let $U(x)$ be the operator defined above. Then

  i. *$U(x)$ defines a representation of the Weyl system $(\mathcal{H}, \mathrm{Im}\,(x,y))$.*
  ii. *This respresentation is faithful and for $t \in \mathbb{R}$ the unitary group $U(tx)$ is strongly continuous.*

Proof: First one shows that $U(x)$ is unitary by the following calculation:

$$(U(y)\exp x', U(y)\exp x)$$

$$= (e^{\{-1/2\|y\|^2-(y,x')\}}\exp(y+x'), e^{\{-1/2\|y\|^2-(y,x)\}}\exp(y+x))$$

$$= e^{\{-\|y\|^2-\overline{(y,x')}-(y,x)\}}e^{(y+x',y+x)} = e^{(x',x)} = (\exp x', \exp x).$$

This means $U(y)$ is isometric, but since $U(y)$ maps a total set onto a total set, it follows that $U(y)$ is unitary. Next one obtains:

$$e^{i\sigma(x,y)}U(x)U(y)\exp z$$

$$= e^{i\sigma(x,y)}e^{\{-1/2\|x\|^2-(x,y+z)-1/2\|y\|^2-(y,z)\}}\exp(x+y+z)$$

$$= e^{\{-1/2\|x+y\|^2-(x+y,z)\}}\exp(x+y+z) = U(x+y)\exp z.$$

This shows that $U(x)$ fulfils the proper algebraic relation.

From the fact that $\exp x$ and $\exp x'$ coincide only if x and x' coincide one sees that $U(x) \neq U(x')$ for $x \neq x'$, and hence one has a faithful representation of the algebra $\mathcal{A}_{\mathcal{H},\sigma}$. Since the expression

$$U(tx)\exp y = e^{\{-\frac{t^2\|x\|^2}{2}-t(x,y)\}}\exp(tx+y) \quad t \in \mathbb{R}$$

is analytic in t it follows that $U(tx)$ is strongly continuous in t. □

It remains to show that the representation on $S\mathcal{H}_m$ defines a respresentation fulfilling the spectrum condition.

I.2.6 Proposition:

Take $\mathcal{H} = \mathcal{H}_m$. Then:

  a. *The linear operators defined by*

$$V(g)\exp \psi = \exp \psi_g, \quad \psi \in \mathcal{H}_m$$

  define a continuous unitary representation of the orthochronous Poincaré group fulfilling the spectrum condition.
  b. *$V(g)$ induces the automorphism α_g on the representation $U(\varphi)$ of the Weyl-system*

$$V(g)U(\varphi)V^{-1}(g) = U(\varphi_g) = \pi(\alpha_g u(\varphi)).$$

Hence $\{\mathcal{A}(O), \mathcal{A}_m, \mathbb{R}^d, \alpha\}$ fulfils all the axioms of section I.1.

Proof: Since the measure $\delta(p^2 - m^2)\Theta(p_0)$ is invariant under orthochronous Poincaré transformation it follows that $v(g)\psi = \psi_g$ defines a unitary representation of the Poincaré group on $\mathcal{H}_m$. Since the translations correspond in momentum space to multiplication by e^{ipa} one sees that $v(a)$ fulfils the spectrum condition. On $\mathcal{H}_m \otimes \mathcal{H}_m \otimes \cdots \otimes \mathcal{H}_m$ the group representation $v(a) \otimes v(a) \cdots \otimes v(a)$ fulfils again the spectrum condition and has $S^n\mathcal{H}_m$ as an invariant subspace. Hence defining

$$V(a)\sum S^n\mathcal{H}_m = \sum v^{\otimes n}(a)S^n\mathcal{H}_m$$

one obtains a continuous unitary representation of the orthochronous Poincaré group fulfilling the spectrum condition and $v(g)\varphi = \varphi_g$.

Now the definition of $U(\varphi)$ gives

$$V(g)U(\varphi)V^{-1}(g)\exp\psi = e^{-1/2\|\varphi\|^2 - (\varphi, \psi_{g^{-1}})}V(g)\exp(\varphi + \psi_{g^{-1}})$$
$$= e^{-1/2\|\varphi_g\|^2 - (\varphi_g, \psi)}\exp(\varphi_g + \psi) = U(\varphi_g)\exp\psi.$$

But this shows that $V(g)$ implements the automorphism α_g. $\qquad\square$

This model corresponds to a scalar free field of mass m. It can be generalized in several directions.

i. One can take the limit $m \to 0$. In the case of one space and one time dimension one has to take care about the point 0 in momentum space. In order that all formulas make sense one has to start with functions vanishing at the origin of the momentum space.

ii. Let μ be a measure on the real line which is polynomially bounded and vanishes for negative values. Then in all the previous considerations $\mathcal{H}_m$ can be replaced by $\int_0^\infty \mathcal{H}_m \, d\mu(m)$. In this case $\Delta_m^+(x)$ is replaced by $\int_0^\infty \Delta_m^+(x) \, d\mu(m) =: \Delta_\mu^+(x)$.

iii. $\Delta_\mu^+(x)$ is again analytic for spacelike x. This remains true if one replaces $\Delta_\mu^+(x)$ by $P(D)\Delta_\mu^+(x)$ where $P(D)$ is any polynomial in the derivatives. However, as one needs a positive scalar product one finds that P is of the form $\widetilde{P}(-iD)$ with $\widetilde{P}(q) \geq 0$ for $q \in \overline{V}^+$. If P is not an invariant polynomial one loses the invariance under Lorentz transformation.

iv. Finally one can replace the "functions" $\Delta^+(x)$ by a matrix $\Delta_{i,k}^+(x)$, $i, k = 1, \ldots, n$ where $\mathcal{F}^{-1}\Delta_{i,k}^+(p) = \nu_{i,k}$ are measures with support in $\overline{V}^+$, and each of the measures is of the form $\nu(p) = P(p)\mu(p)$ with $\mu(p)$ an invariant measure. Moreover, for every $(c_i) \in \mathbb{C}^n$ the measure $\sum_{i,j} \overline{c_i}c_j\nu_{i,j}(p)$ has to be positive.

The last extension allows one to describe also the so-called fields with integer spin. All these examples are known under the name of generalized free fields.

I.3   Examples: Free Fermi fields

Another family of examples fulfilling the axioms of section I.1 can be extracted from the canonical anti-commutation relations (C.A.R.). Again we start with a short discussion of the C.A.R. which also appear in different forms.

Case I:

This is called the Clifford algebra case. Let E be a real Hilbert space with real non-degenerate scalar product. One is looking for a linear mapping φ of the space E into a C^*–algebra satisfying the anti-commutation relation

$$\{\varphi(f), \varphi(g)\} - 2(f, g)\mathbb{1} = 0$$

with $\{\varphi, \chi\} = \varphi\chi + \chi\varphi$ and the condition $\varphi(f) = \varphi(f)^*$. From these relations one concludes $\|\{\varphi(f), \varphi(f)\}\| = 2\|\varphi(f)^2\| = 2(f, f)$ so that φ is an isometric map. The Clifford algebra is the free algebra generated by the elements in $\mathcal{H}$ modulo the two-sided ideal generated by the elements of the form $\{\varphi(f), \varphi(g)\} - 2(f, g)\mathbb{1}$. Then φ is a representation of the Clifford algebra.

Case II:

This is usually called the Fermi case. Let E again be a real Hilbert space. Now one wants to associate to every $f \in E$ two operators $\varphi(f)$ and $\varphi(f)^*$ such that $\varphi(f)^*$ is the adjoint of $\varphi(f)$ and both fulfil the anti-commutation relations

$$\{\varphi(f), \varphi(g)\} = \{\varphi(f)^*, \varphi(g)^*\} = 0$$

and

$$\{\varphi(f), \varphi(g)^*\} = (f, g)\mathbb{1}.$$

For computing the norm of $\varphi(f)$ notice first that $\varphi(f)^2 = \frac{1}{2}\{\varphi(f), \varphi(f)\} = 0$. So one obtains:

$$\varphi(f)^*\varphi(f)\varphi(f)^*\varphi(f) = \varphi(f)^*\{\varphi(f), \varphi(f)^*\}\varphi(f)$$
$$= (f, f)\varphi(f)^*\varphi(f).$$

Consequently, if $\|.\|$ is a C^*-norm

$$\|\varphi^*(f)\varphi(f)\|^2 = (f, f)\|\varphi(f)^*\varphi(f)\|$$

and if $\varphi(f) \neq 0$

$$\|\varphi(f)\|^2 = (f, f) = \|f\|^2,$$

which implies that the map φ is again isometric.

Case III:

This is the Fermi-Dirac case. Let $\mathcal{H}$ be a complex Hilbert space and let K be a conjugation on $\mathcal{H}$. By a conjugation we mean an antiunitary operator K on $\mathcal{H}$ which fulfills $K^2 = \mathbb{1}$ and $K^* = K$. Define:

$$\widehat{\mathcal{H}} = \mathcal{H} \oplus \mathcal{H}$$

and

$$\widehat{K} = \begin{pmatrix} 0 & K \\ K & 0 \end{pmatrix}.$$

$\widehat{K}$ is a conjugation on $\widehat{\mathcal{H}}$. Define on $\widehat{\mathcal{H}}$ a bilinear form

$$< \widehat{f}, \widehat{g} > = (\widehat{K}\widehat{f}, \widehat{g}) = (Kf_2, g_1) + (Kf_1, g_2)$$

with $\widehat{f} = f_1 \oplus f_2, \widehat{g} = g_1 \oplus g_2$.

Now one looks for a complex linear mapping ψ of $\widehat{\mathcal{H}}$ into a C^*-algebra satisfying the relations

$$\{\psi(\widehat{f}), \psi(\widehat{g})\} = < \widehat{f}, \widehat{g} > \mathbb{1}$$

and

$$\psi(\widehat{f})^* = \psi(\widehat{K}\widehat{f}).$$

Let $\mathbf{T}$ be the one-dimensional torus. Define for $\widehat{f} \in \widehat{\mathcal{H}}$ a map β_φ by the equation

$$\beta_\varphi(f_1 \oplus f_2) = e^{i\varphi} f_1 \oplus e^{-i\varphi} f_2.$$

β_φ is an isometry of $\widehat{\mathcal{H}}$. Moreover,

$$\begin{aligned} < \beta_\varphi \widehat{f}, \beta_\varphi \widehat{g} > &= (Ke^{-i\varphi} f_2, e^{i\varphi} g_1) + (Ke^{i\varphi} f_1, e^{-i\varphi} g_2) \\ &= (Kf_2, g_1) + (Kf_1, g_2) = < \widehat{f}, \widehat{g} > . \end{aligned}$$

This shows that the bilinear form is invariant under the transformation β_φ. This implies together with $\beta_\varphi \widehat{K} = \widehat{K} \beta_\varphi$ that the equation

$$\alpha_\varphi \psi(\widehat{f}) = \psi(\beta_\varphi \widehat{f})$$

defines an automorphism of the C^*-algebra $\mathcal{A}(\widehat{\mathcal{H}})$ generated by the $\psi(\widehat{f})$. Moreover, α_φ is strongly continuous on polynomials in ψ and since these polynomials are norm-dense in $\mathcal{A}(\widehat{\mathcal{H}})$ it follows that α_φ defines a strongly continuous group of automorphisms on $\mathcal{A}(\widehat{\mathcal{H}})$.

Since $\mathbf{T}$ is a compact group with dual $\mathbb{Z}$ every $A \in \mathcal{A}(\widehat{\mathcal{H}})$ can be written as a sum

$$A = \sum_{-\infty}^{\infty} A_n \quad \text{with} \quad \alpha_\varphi A_n = e^{in\varphi} A_n.$$

Set
$$\mathcal{A}_n(\widehat{\mathcal{H}}) = \{A \in \mathcal{A}(\widehat{\mathcal{H}}); \alpha_\varphi A = e^{in\varphi} A\}.$$

Then
$$\mathcal{A}(\widehat{\mathcal{H}}) = \sum_{-\infty}^{\infty} \mathcal{A}_n(\widehat{\mathcal{H}})$$

where $\mathcal{A}_n(\widehat{\mathcal{H}})$ are norm-closed linear subspaces of $\mathcal{A}(\widehat{\mathcal{H}})$ satisfying the multiplication rule
$$\mathcal{A}_n(\widehat{\mathcal{H}})\mathcal{A}_m(\widehat{\mathcal{H}}) \subseteq \mathcal{A}_{n+m}(\widehat{\mathcal{H}})$$

with
$$\{\mathcal{A}_n(\widehat{\mathcal{H}})\}^* = \mathcal{A}_{-n}(\widehat{\mathcal{H}}).$$

This shows that the Fermi-Dirac algebra $\mathcal{A}(\widehat{\mathcal{H}})$ is a $\mathbf{Z}$-graded C^*–algebra.

For constructing a free field describing charged particles of spin 1/2, we choose 4×4 matrices γ^i, $i = 0, \ldots, 3$ fulfilling the relations
$$\gamma^i\gamma^k + \gamma^k\gamma^i = 2g^{i,k}.$$

Since $g^{00} = 1$ and $g^{ii} = -1$, $i = 1, 2, 3$ one can choose γ in such a way that
$$(\gamma^0)^* = \gamma^0, \quad (\gamma^i)^* = -\gamma^i \quad i = 1, 2, 3.$$

With these matrices, using the summation convention, the Dirac equation reads:
$$(\frac{1}{i}\gamma^j\partial_j - m)\psi(x) = 0$$

where $\psi(x)$ is a four component vector function. In momentum space the Dirac equation reads:
$$(\gamma^j p_j - m)\psi(p) = 0.$$

Applying to this equation the operator $(\gamma^k p_k + m)$ one obtains as a consequence of the anti-commutation relation for the γ's the equation
$$(p^2 - m^2)\psi(p) = 0.$$

This implies that the solutions of the Dirac equation are also solutions of the wave equation.

Introducing the operators
$$Q^\pm(p) = \frac{1}{2m}(m \pm \gamma^i p_i),$$

a simple calculation gives the following relations:

$$Q^+(p)Q^+(p) = Q^+(p) + \frac{1}{4m^2}(p^2 - m^2)$$

$$Q^-(p)Q^-(p) = Q^-(p) + \frac{1}{4m^2}(p^2 - m^2)$$

$$Q^+(p)Q^-(p) = Q^-(p)Q^+(p) = \frac{-1}{4m^2}(p^2 - m^2)$$

$$Q^+(p) + Q^-(p) = 1.$$

This shows that these operators are projections when restricted to the hyperboloid $p^2 = m^2$.

Since the Dirac operator is nothing else but $-2mQ^-(p)$ it follows that the restriction of $Q^+(p)\psi(p)$ to $p^2 = m^2$ is a solution of the Dirac equation. This property remains true if one multiplies Q^+ by a scalar λ which is allowed to depend on p. A special role in the quantization of the Dirac field is played by the operator

$$R = \sqrt{\frac{m}{|p_0|}}Q^+.$$

In order to understand this factor let us try to define a Hilbert space of solutions of the Dirac equation. A first attempt might be

$$(g, f)_1 = \int (Q^+(p)g(p), Q^+(p)f(p))\delta(p^2 - m^2)dp$$

with

$$(Q^+(p)g(p), Q^+(p)f(p)) = \bar{g}_\nu(p)(Q^{+*}Q^+)_{\nu,\mu}f_\mu(p).$$

For the computation of the expression $Q^{+*}Q^+$ use the above–mentioned choice $\gamma^{0*} = \gamma^0, \gamma^* = -\gamma$, to obtain:

$$Q^+(p) = \frac{1}{2m}(m + \gamma^0 p_0 + (\underline{\gamma}, \mathbf{p})) \quad \text{and}$$

$$Q^+(p)^* = \frac{1}{2m}(m + \gamma^0 p_0 - (\underline{\gamma}, \mathbf{p})) \quad \text{with} \quad (\underline{\gamma}, \mathbf{p}) = \sum_1^3 \gamma^l p_l.$$

Using the anti-commutation relation for the γ's one obtains:

$$Q^+(p)^*Q^+(p) = \frac{1}{4m^2}(m^2 + 2m\gamma^0 p_0 + {\gamma^0}^2 p_0^2 + \gamma^0 p_0(\underline{\gamma}, \mathbf{p})$$
$$- (\underline{\gamma}, \mathbf{p})\gamma^0 p_0 - (\underline{\gamma}, \mathbf{p})^2)$$
$$= \frac{1}{4m^2}(m^2 - p^2 + 2\gamma^0 p_0(m + \gamma^i p_i))$$
$$= \frac{1}{4m^2}(m^2 - p^2) + \frac{p_0}{m}\gamma^0 Q^+.$$

The first term vanishes when restricted to the mass hyperboloid. The second term contains p_0 as factor, but $p_0\delta(p^2 - m^2)$ is not local in configuration space. Therefore we have to look for another scalar product.

It is our intention to use the Fermi-Dirac method described under Case III to obtain a quantized Dirac field having vanishing anti–commutators for spacelike separations. This means a local function should appear in scalar product. But the Fourier transform of $\delta(p^2 - m^2)$ is not a local function, however the Fourier transform of $\epsilon(p_0)\delta(p^2-m^2)$ is $(\epsilon(\lambda) = \text{sign } \lambda)$. Therefore if one replaces Q^+ by $R = \sqrt{\frac{m}{|p_0|}}Q^+$ one obtains:

$$R^*R = \gamma^0 Q^+ \epsilon(p_0) + \frac{1}{4m|p_0|}(p^2 - m^2).$$

Finally one may define the scalar product as follows:

$$(g, f) = \int (R(p)g(p), R(p)f(p))\delta(p^2 - m^2)\,dp$$

$$= \int (g(p), \gamma^0 Q^+ f(p))\epsilon(p_0)\delta(p^2 - m^2)\,dp$$

$$= \int (g(x), \gamma^0 S(x - y)f(y))\,dx\,dy$$

where $S(x)$ is the Fourier transformation of

$$Q^+(p)\epsilon(p_0)\delta(p^2 - m^2).$$

Notice $\Delta(x) = \mathcal{F}(\epsilon(p_0)\delta(p^2 - m^2))$ vanishes for spacelike x, as does

$$S_m(x) = \frac{1}{2m}\left(m + \frac{1}{i}\gamma^j \partial_j\right)\Delta_m(x).$$

Now we are ready to introduce the *field algebra* associated with the Hilbert space of solutions of the Dirac equation.

I.3.1 Definition:

Let $\mathcal{H}$ be the Hilbert space of solutions of the Dirac equation with the above scalar product and let K be the complex conjugate in configuration space

$$Kf(x) = \overline{f(x)}.$$

(a) *Define $\mathcal{F}(O)$ to be the smallest C^*–algebra generated by the Fermi-Dirac operators $\psi(\widehat{f})$, $\widehat{f} \in S(\mathbb{R}^4) \times (\mathbb{C}^4 \oplus \mathbb{C}^4)$ and supp $\widehat{f} \subset O$.*

(b) *$\mathcal{F}$ is defined as the C^*–inductive limit of all the $\mathcal{F}(O)$ and it is called the field algebra.*

(c) *Define a translation by the relation*

$$\alpha_a\psi(\widehat{f}) = \psi(\widehat{f_a}), \quad a \in \mathbb{R}^4.$$

This extends to an automorphism of $\mathcal{F}$ since the map $\widehat{f} \to \widehat{f_a}$ is a unitary transformation of $\widehat{\mathcal{H}}$ leaving invariant the bilinear form $<\widehat{f}, \widehat{g}>$.

(d) *If $\widehat{f}$ is of the form $f \oplus 0$ then write $\psi(f)$ instead of $\psi(\widehat{f})$ and if $\widehat{g}$ has the form $0 \oplus g$ then write $\psi^*(g)$ instead of $\psi(\widehat{g})$. This notation is justified by the relation*

$$(\psi(f))^* = \psi^*(\overline{f}).$$

(e) *Let M be a monomial in the operators $\psi(f)$ and $\psi^*(g)$. Let $n(M)$ denote the number of factors of $\psi(f)$ and $n^*(M)$ that of $\psi^*(g)$ appearing in M, so that $n(M) + n^*(M)$ denotes the number of factors in M.*

With this notation one obtains:

I.3.2 Lemma:

(1) *$\mathcal{F}(O)$ is a graded algebra*

$$\mathcal{F}(O) = \sum_{-\infty}^{+\infty} \mathcal{F}_i(O).$$

(2) *A monomial $M \in \mathcal{F}(O)$ belongs to $\mathcal{F}_i(O)$ if and only if*

$$n(M) - n^*(M) = i.$$

These monomials are total in $\mathcal{F}_i(O)$.

(3) *Let O_1 and O_2 be spacelike separated and assume $A \in \mathcal{F}_i(O_1), B \in \mathcal{F}_k(O_2)$ then one has the commutation relation*

$$AB = (-1)^{ik} BA.$$

Proof: Statement (1) is a consequence of the discussion of the third case of this section.

(2) From the definition of ψ and ψ^* one obtains $\alpha_\varphi \psi(f) = e^{i\varphi}\psi(f)$ and $\alpha_\varphi \psi^*(g) = e^{-i\varphi}\psi^*(g)$. Since α_φ is an automorphism one gets for a monomial the equation

$$\alpha_\varphi M = \exp\{i(n(M) - n^*(M))\varphi\}M,$$

and hence $M \in \mathcal{F}_i(O)$ if $i = n(M) - n^*(M)$. Since one can read this equation in both directions, the first part of (2) follows. Since the polynomials are norm-dense in the field algebra it follows, by projecting onto $\mathcal{F}_i$, that the monomials are total in $\mathcal{F}_i$.

(3) The definition of ψ and ψ^* implies that

$$\{\psi(f), \psi(g)\} = \{\psi^*(f), \psi^*(g)\} = 0, \quad \text{and}$$

$$\{\psi(f), \psi^*(g)\} = (Kf, g) = \int \sum_{\mu, \nu} f_\nu (\gamma^0 S(x - y))_{\nu, \mu} g_\mu \, dx \, dy.$$

Hence $\psi(f)$ and $\psi^*(g)$ anti-commute when supp $f \subset O_1$, supp $g \subset O_2$ and these sets are spacelike separated. If $M_1 \in \mathcal{F}(O_1)$ and $M_2 \in \mathcal{F}(O_2)$ then one obtains:

$$M_1 M_2 = (-1)^{(n(M_1)+n^*(M_1))(n(M_2)+n^*(M_2))} M_2 M_1$$
$$= (-1)^{(n(M_1)-n^*(M_1))(n(M_2)-n^*(M_2))} M_2 M_1.$$

From this one obtains (3) by the density proved in (2). □

For constructing examples of local nets the following notation is helpful:

I.3.3 Notation:

(a) *By $\mathcal{F}^n(O)$ we denote the subalgebra of the form*

$$\mathcal{F}^n(O) = \sum_{i=-\infty}^{+\infty} \mathcal{F}_{i \cdot n}(O)$$

and

$$\mathcal{F}^0(O) = \mathcal{F}_0(O).$$

(b) *$\mathcal{F}^n$ and $\mathcal{F}^0$ denote the corresponding objects in the C^*-inductive limit.*
(c) *Denote by $\mathbf{T}$ the group of all α_φ and by $\mathbb{Z}_n$ the subgroup generated by the automorphism $\alpha_{\frac{2\pi}{n}}$, $n \neq 0$.*
(d) *For $a \in \mathbb{R}^4$ define*

$$\alpha_a \psi(f) = \psi(f_a)$$

and

$$\alpha_a \psi^*(f) = \psi^*(f_a).$$

Since the scalar product of $\mathcal{H}$ is invariant under the map $f \to f_a$ it follows that α_a extends to an automorphism of the Fermi-Dirac algebra.

With this notation one obtains the following results:

I.3.4: Lemma

(1) *An element $A \in \mathcal{F}$ belongs to the subalgebra $\mathcal{F}^n$ if and only if A is invariant under α_g, $g \in \mathbb{Z}_n$ i.e.,*

$$\alpha_g A = A, \quad g \in \mathbb{Z}_n, \quad n \neq 0$$

and $A \in \mathcal{F}^0$ if and only if $\alpha_\varphi A = A$, $\forall \varphi \in \mathbf{T}$.
(2) *The system $\{\mathcal{F}^{2n}(O), \mathcal{F}^{2n}, \alpha, \mathbb{R}^4\}$ fulfils the axioms I, II, and III for $n = 0, 1, 2, \ldots$*

Remark:

It is also possible to define automorphisms for every Poincaré tranformation. Therefore the above systems also fulfil the Axiom III P. But the details will not be discussed here.

Proof: (1) If $A \in \mathcal{F}_j$ then one has $\alpha_\varphi A = e^{ij\varphi}A$. This implies if $A \in \mathcal{F}^n$ and $\varphi \in Z_n$ the relation $\alpha_\varphi A = A$. Conversely if $A \in \mathcal{F}$ and $\alpha_{\frac{k2\pi}{n}} A = A$, then with $A = \sum_{-\infty}^{+\infty} A_j$ one has

$$\alpha_{\frac{k2\pi}{n}} A = \sum_j \exp\{ij\frac{k2\pi}{n}\}A_j = \sum_{m=0}^{n-1} \exp\{i\frac{m2\pi}{n}\} \sum_{l=-\infty}^{+\infty} A_{ln+m}.$$

Hence $\sum_{l=-\infty}^{+\infty} A_{ln+m} = 0$ for $m = 1, ...n-1$ which implies $A = \sum_p A_{pn} \in \mathcal{F}^n$. This argument remains true for $n = 0$ if one takes all α_φ.

(2) $2n$ is even. Hence one obtains from Lemma I.3.2(3) that $\mathcal{F}^{2n}(O_1)$ and $\mathcal{F}^{2n}(O_2)$ commute with each other if O_1 and O_2 are spacelike separated. Since, moreover, $f \to f_a$ has the right transformation properties it follows that the system $\{\mathcal{F}^{2n}(O), \mathcal{F}^{2n}, \alpha_a, \mathbf{R}^4\}$ fulfils the axioms I, II, and III. □

In order to show that there are also examples for the theory of local observables it remains to construct a faithful representation fulfilling the spectrum condition.

I.3.5 Proposition:

The theories $\{\mathcal{F}^{2n}(O), \mathcal{F}^{2n}, \alpha, \mathbf{R}^4\}$, $n = 0, 1...$ admit faithful representations fulfilling the spectrum condition, and hence they define examples of the theory of local observables.

Proof: Since $\mathcal{F}^{2n}$ is a subalgebra of $\mathcal{F}$ it is sufficient to construct a faithful representation of $\mathcal{F}$ fulfilling the spectrum condition. We will use a momentum space respresentation. Let us remark first that K is local in configuration space, which means $Kf(x) = \overline{f(x)}$. This implies that $Kf(p) = \overline{f(-p)}$ for momentum space.

Denote by $\mathcal{H}_m$ the Hilbert space defined by the scalar product.

$$(f, g) = \int \sum_{\nu,\mu} \overline{f_\nu(p)}(R^*R)_{\nu,\mu}(p)g_\mu(p)\Theta(p_0)\delta(p^2 - m^2)\, dp.$$

Introduce a second Hilbert space $\mathcal{H}_m^t$ by the scalar product

$$(f, g)_t = \int \sum_{\nu,\mu} \overline{f_\nu(p)}(R^*R)_{\nu,\mu}^t g_\mu(p)\Theta(p_0)\delta(p^2 - m^2)\, dp$$

$$= \int \sum_{\nu,\mu} \overline{f_\nu(p)}(R^*R)_{\mu,\nu}g_\mu(p)\Theta(p_0)\delta(p^2 - m^2)\, dp$$

where $(R^*R)^t$ denotes the transpose of the matrix (R^*R). On $\mathcal{H}_m$ and on $\mathcal{H}_m^t$ there exist natural unitary representations of the translations $u(a)$ and $v(a)$ defined by:

$$u(a)f(p) = e^{ipa}f(p), \quad v(a)f(p) = e^{ipa}f(p).$$

Both these representations are unitary and fulfil the spectrum condition. By $A^n \mathcal{H}_m$ and $A^n \mathcal{H}_m^t$ we denote the anti–symmetric part of the n-th tensorial power of $\mathcal{H}_m$ and of $\mathcal{H}_m^t$ respectively. One finally defines a Hilbert space $\mathcal{H}$ as follows:

$$\mathcal{H} = \sum_{k,l=0}^{\infty} A^k \mathcal{H}_m \otimes A^l \mathcal{H}_m^t.$$

Since anti-symmetrization commutes with the translations the definition

$$U(a) A^k \mathcal{H}_m \otimes A^l \mathcal{H}_m^t = \otimes^k u(a) \otimes^l v(a) A^k \mathcal{H}_m \otimes A^l \mathcal{H}_m^t$$

gives a unitary representation acting on $\mathcal{H}$ which fulfils the spectrum condition.

If a function $\Phi(p_1, ..., p_k; q_1, ..., q_l)$ belongs to $A^k \mathcal{H}_m \otimes A^l \mathcal{H}_m^t$ then it is anti-symmetric in the first variables $p_1, ..., p_k$ and also anti-symmetric in the variables $q_1, ..., q_l$.

Now one defines operators ψ and ψ^* as follows:

$$\psi(f)\Phi(p_1, ..., p_k; q_1, ...q_l)$$
$$= \frac{1}{\sqrt{k+1}} \sum_{1}^{k+1} (-1)^{j+1} f(p_j) \Phi(p_1, ..., \widehat{p}_j, ..., p_{k+1}; q_1, ..., q_l)$$
$$+ (-1)^k \sqrt{l} \left(\overline{f(-q_1)}, \Phi(p_1, ...p_k; q_1, ..., q_l) \right)_t$$

and

$$\psi^*(g)\Phi(p_1, ..., p_k; q_1, ..., q_l)$$
$$= \sqrt{k} \left(\overline{g(-p_1)}, \Phi(p_1, ...p_k; q_1, ..., q_l) \right)$$
$$+ (-1)^k \frac{1}{\sqrt{l+1}} \sum_{1}^{l+1} (-1)^{j+1} g(q_j) \Phi(p_1, ...p_k; q_1, ..., \widehat{q}_j, ...q_l),$$

where the symbol $\widehat{p}_j$ means that the variable p_j has been dropped and the "scalar product" the integration over the corresponding variable.

It is easy to check that these operators fulfil the relations

$$\psi(f)^* = \psi^*(Kf), \quad U(a)\psi(f)U^*(a) = \psi(f_a),$$

so that it remains to show that these operators also satisfy the right anti-commutation relations. Using the above definitions one obtains:

$$(\psi^*(g)\psi(f)\Phi(p_1,...p_k;q_1,...,q_l)$$

$$= \frac{\sqrt{k+1}}{\sqrt{k+1}}\left(\overline{g(-p_1)}, \sum_j (-1)^{j+1}f(p_j)\Phi(p_1,...,\widehat{p}_j,...,p_{k+1};q_1,...,q_l)\right)$$

$$+ \frac{1}{\sqrt{k+1}\sqrt{l+1}}$$

$$\sum_{i,j}(-1)^{i+j+2}f(p_i)g(q_j)\Phi(p_1,...,\widehat{p}_i,...,p_{k+1};q_1,...,\widehat{q}_j,...,q_{l+1})$$

$$+ (-1)^k\sqrt{k}\sqrt{l}\left(\overline{g(-p_1)}, \left(\overline{f(-q_1)}, \Phi(p_1,...p_k;q_1,...,q_l)\right)_t\right)$$

$$+ (-1)^{k+k}\frac{\sqrt{l}}{\sqrt{l}}$$

$$\sum_j (-1)^{j+1}g(q_j)\left(\overline{f(-q_o)}, \Phi(p_1,...p_k;q_o,q_1,...,\widehat{q}_j,...,q_l)\right)_t$$

and for the reversed order:

$$\psi(f)\psi^*(g)\Phi(p_1,...p_k;q_1,...,q_l)$$

$$+ \frac{\sqrt{k}}{\sqrt{k}}\sum_j (-1)^{j+1}f(p_j)\left(\overline{g(p_o)}, \Phi(p_o,p_1,...\widehat{p}_j,...,p_k;q_1,...,q_l)\right)$$

$$+ (-1)^{k+1}\sqrt{k}\sqrt{l}\left(\overline{f(-q_1)}, \left(\overline{g(-p_1)}, \Phi(p_1,...p_k;q_1,...,q_l)\right)\right)_t$$

$$+ \frac{(-1)^k}{\sqrt{k+1}\sqrt{l+1}}$$

$$\sum_{i,j}(-1)^{i+j+2}f(p_i)g(q_j)\Phi(p_1,...,\widehat{p}_i,...,p_{k+1};q_1,...,\widehat{q}_j,...,q_{l+1})$$

$$+ (-1)^{k+k}\frac{\sqrt{k+1}}{\sqrt{l+1}}$$

$$\left(\overline{f(-q_1)}, \sum_j (-1)j+1 g(q_j)\Phi(p_1,...p_k;q_1,...,\widehat{q}_j,...,q_{l+1})\right)_t.$$

Taking the sum we obtain:

$$(\psi^*(g)\psi(f) + \psi(f)\psi^*(g))\Phi(p_1,...p_k;q_1,...,q_l)$$

$$= \left(\overline{g(-p_1)}, f(p_1)\right) + \left(\overline{f(-q_1)}, g(q_1)\right)_t \Phi(p_1,...p_k;q_1,...,q_l)$$

$$= \int \{g_{nu}(-p)(R^*R)_{\nu,\mu}f_\mu(p) + f_{nu}(-p)(R^*R)_{\mu,\nu}g_\mu(p)\}$$

$$\Theta(p_0)\delta(P^2 - m^2)\,dp\,\Phi(p_1,...p_k;q_1,...,q_l)$$

$$= \int g_\nu(-p)(R^*R)_{\nu,\mu}f_\mu(p)\delta(p^2 - m^2)\,dp\,\Phi(p_1,...p_k;q_1,...,q_l).$$

For the last step we have used the transformation $p \to -p$.

The relation $\{\psi(f), \psi(g)\} = 0$ can be proved in the same manner. But this proves the proposition since the representation ψ and ψ^* is obviously faithful. □

I.4 Notes and remarks

Quantum field theory dates back to the late twenties and early thirties when Dirac [Dir27] tried to quantize the electromagnetic field and when Jordan and Wigner [JW] gave a description of the free quantized Fermi gas. However, it was soon realized that these concepts could not easily be generalized to interacting theories. After a period of twenty years the first axiomatic approach to field theory was constructed by A.S. Wightman [Wi56]. Again ten years later the algebraic structure of the Wightman approach was revealed by the author [Bch62] and independently by Uhlmann [Uhl]. This theory uses unbounded operators as basic objects. In classical quantum mechanics many of the observables are given by self–adjoint operators and it is often more useful to investigate the bounded functions of the observables than the unbounded ones. In this spirit Haag started at about the same time to develop a quantum field theory in terms of bounded operators. His ideas are presented in [Haa]. A first consistent presentation of this subject was given by H. Araki [Ar61] in a series of lectures held at the E.T.H. in Zürich during the academic year 1961/62. A discussion of the axioms has been given by Haag and Schroer [HS]. A modern introduction to this subject can be foung in the book of R. Haag [[Haa]].

A large group of people working in this direction hoped that quantum field theory could be formulated entirely in terms of C^*–algebras. It is clear that a C^*–algebra fulfilling all axioms except that of the spectrum condition must have an enormous number of inequivalent representations. D. Kastler had the idea that Fell's equivalence [Fe] of representations leads to reductions of this large number of inequivalent classes. A detailed discussion of this phenomenon and its physical meaning has been given by Haag and Kastler [HK].

The axioms formulated here are not completely algebraic. The spectrum condition has to refer to representations. But one cannot get a satisfactory theory without the spectrum condition. Indeed Doplicher, Regge and Singer [DRS] have constructed a theory fulfilling the locality condition, which however does not have any representation fulfilling the spectrum condition. If one is willing to assume that the translations are strongly continuous and that all vacuum states together separate the algebra one can give an algebraic version of the spectrum condition. Results in this direction have been formulated by S. Doplicher [Dop] and also by K. Kraus [Kr]. Necessary and sufficient conditions for the existence of a faithful representation fulfilling the spectrum condition have been given by the author [Bch70a]

It is believed that the framework of Wightman and the framework of Araki, Haag and Kastler are mathematically equivalent. The first attempt in this direction was made by W. Zimmermann and the author [BZ] showing that this passage is possible when additional requirements are made on the vacuum state. Recently J. Yngvason and J. Alcantara [AY] took up this question again. They gave necessary and sufficient conditions on a Wightman state which allow the construction of an Araki-Haag-Kastler theory. The converse problem, constructing Wightman field from a theory of local observables has been investigated by Fredenhagen and Hertel [FH]. For a review of the situation concerning the equivalence of the Wightman and Araki–Haag–Kastler framework see the report by J. Yngvason [Y] and the paper by the author and J. Yngvason [BY].

The integral representation of continuous unitary representation of abelian groups which has been used for the discussion of the spectrum condition is due to M. Stone [Sto].

By defining observables as the set of self–adjoint elements of a C^* algebra, which is a Jordan algebra, the concept of symmetry leads to Jordan automorphisms. They can be either automorphisms or an antiautomorphisms, or of a mixed type. Antiautomorphisms are sometimes useful for discrete symmetries as for instance time reversal invariance. But for continuous groups the use of Jordan automorphisms, which are not automorphisms, seems rather artificial. For a discussion see H. Roos [Roo] and references therein.

The canonical commutation relations for a finite number of degrees of freedom were first been used by Heisenberg [Hei]. The explicit form

$$[p_i, q_j] = -i\hbar, \quad [q_i, q_j] = 0 = [q_i, q_j]$$

appeared for the first time in a paper by M. Born and Jordan [BJ]. The integrated form which we have used here is due to H. Weyl [We]. Quantization of an infinite number of degrees of freedom was used the first time by Dirac [Dir27] for the electric field. The algebra used here has been introduced by J. Slawny [Sl] and later but independently by Manuceau, Sirugue, Testard, and Verbeure [MSTV]. There are different ways of associating C^*–algebras with the C.C.R. Most of them have been described in the book by A. Guichardet [[Gui]]. All these algebras have the disadvantage that, in the case of infinitely many degrees of freedom, they also give rise to non-continuous representations of the C.C.R. At present it is not known whether or not there exists, also for this case, a C^*–algebra the representations of which are in one to one correspondence with the continuous representations of the C.C.R. (continuity refers to the continuity of $U(tx)$ in $t \in \mathbb{R}$). In case that the symplectic space is finite and hence $2n$ dimensional D. Kastler [Kas] has constructed a C^*–algebra whose representations are in one to one correspondence with the continuous representation of the Weyl relations. He starts from $L^1(E)$ and defines a new product by

$$(fg)(x) = \int e^{-i\sigma(x,y)} g(x-y) f(y) \, dy$$
$$f^*(x) = \overline{f(x)}.$$

The enveloping C^*–algebra of this algebra then is the algebra in question.

Since this works up to now only for the finite dimensional case it is not a great achievement, as all continuous representation of the finite dimensional Weyl relations are quasi-equivalent. This result has been shown by J. v. Neumann [Neu]. If E is infinite dimensional there exist representations which are not quasi-equivalent. Such examples were first given by K.O. Friedrichs [[Fri]].

If one studies a continuous representation $U(tx)$ of the Weyl system (E, σ) then $U(tx)$ is continuous in t and unitary. Hence there exist generators $A(x)$ with $U(tx) = \exp\{itA(x)\}$. These operators are mostly investigated in the physical literature. However, this has to be done with extreme care since these operators cannot be bounded operators. For a discussion of this approach see for example the book of G.G. Emch [[Em]]. Under suitable assumptions the two methods namely using the $A(x)$ and using the Weyl operators are equivalent. The most general result in this direction appears to my knowledge in a paper by Hegerfeldt [Heg].

The first free field which was constructed was the non–interacting electromagnetic field introduced by Dirac [Dir27]. However, the massive free scalar field used here is simpler. The Hall-Wightman theorem used for the describing properties of $\Delta_m^+(x)$ can be found in [HW]. The construction of a positive energy representation used here is due to V. Fock [Fo]. However, the elegant method of using the "exponential" vector is introduced by H. Araki and J. Woods [AWo]. Generalized free fields not obeying the Klein Gordon equation were introduced by O.W. Greenberg [Gr].

The appearence of C.A.R. in physics is implied by the Pauli exclusion principle which holds for particles with half integer spin. The first quantum mechanical description of a spin $1/2$ particle was given by W. Pauli [Pau]. However, his equations are not invariant under relativistic transformations. Such a covariant equation was found by P.A.M. Dirac [Dir28]. The first description of a system of infinitely many degrees of freedom was the "Fermi-Gas" given by Jordan and Wigner [JW].

The Clifford algebra has been investigated extensively by Shale and Steinspring [SS]. The formulation of case III is similar to that of Araki and Wyss

[AWy]. The main point of this description is to show explicitly the existence of "gauge" transformations and the graded structure of the algebra which goes with it. The literature about the C.A.R. algebra is immense and again the reader should refer to the two books [[Gui]] and [[Em]] for more information. The C.A.R. algebra has played an important role in the development of the theory of von Neumann algebras. Using these algebras R.T. Powers [Pow] was able to show for the first time existence of a continuous family of inequivalent type III factors.

For more details on the γ matrices we refer to the textbook of Jauch and Rohrlich [[JR]]. In textbooks treating the quantized Dirac field the quantity $\psi^*(f)$ is usually replaced by $\psi^\dagger(f) := \psi^*(\gamma^0 f)$. Using these operators the anti-commutation relations become

$$\{\psi(f), \psi^\dagger(g)\} = \int f_v(x) S_{v\mu}(x - y) g_\mu(y) \mathrm{d}x dy$$

The gauge transformations $\psi(f) \to e^{i\varphi}\psi(f)$ are closely related to the existence of a charge in this theory, and they are connected with the doubling of the Hilbert space which appears in case III. It should be mentioned that one also can construct "neutral" spin 1/2 fields. This procedure is due to Majorana [Maj].

Chapter II

Translations and the Spectrum Condition

In this chapter we will investigate C^*–dynamical systems $\{\mathcal{A}, \mathbb{R}^d, \alpha\}$, i.e., systems for which the symmetry group is the d-dimensional translation group. Let $\mathbb{R}'^d$ be the dual space of $\mathbb{R}^d$ and $C' \subset \mathbb{R}'^d$ be a proper, closed convex cone with interior points. We want to characterize representations $\{\mathcal{H}, \pi, U, C'\}$ of the above C^*–dynamical system having the properties:

(α) π is a non-degenerate representation of $\mathcal{A}$ acting on $\mathcal{H}$,

(β) $U(a), a \in \mathbb{R}^d$ is a continuous unitary representation of the translation group $\mathbb{R}^d$ implementing the automorphisms α_a, i.e.,

$$U(a)\pi(x)U^*(a) = \pi(\alpha_a x), \quad x \in \mathcal{A}, \quad a \in \mathbb{R}^d.$$

(γ) The spectrum of $U(a)$ is contained in C'.

No special properties of the C^*-Algebra $\mathcal{A}$ will be required in this chapter. In the first section we collect some results about the relations between support properties and analytic properties of Fourier transformations. We shall work within the framework of the theory of tempered distributions. In section two we will investigate the relations between functionals φ on $\mathcal{A}$ and their absolute values. In particular we want to show that the map $\varphi \to |\varphi|$ is continuous in the norm-topology. In section three we introduce the spaces of momentum transfer. They are also called spectral subspaces in the mathematical literature. Having investigated the properties of the spaces of momentum transfer we shall deal in section four with C^*–dynamical systems where the group in question is one-dimensional. In section five, necessary and sufficient conditions for representations with spectrum condition for the one-dimensional case will be derived. Section six handles the same situation as before, but with the additional assumption that the one-parameter group for which the spectrum condition is required is imbedded in a larger group. Finally, in section seven the general case will be treated; this is the case where the group is $\mathbb{R}^d$ and the spectrum is contained in a convex cone C'. For any set G we denote its closure by $\overline{G}$ and its interior points by $\mathring{G}$.

II.1 Relations between support properties of functions and analytic properties of their Fourier transforms

The framework for this investigation will be the theory of tempered distributions. $\mathcal{S}'(\mathbb{R}^d)$ denotes the space of tempered distributions on $\mathbb{R}^d$ and $\mathcal{S}(\mathbb{R}^d)$ denotes correspondingly the space of strongly decreasing test functions. We start with the following notation:

II.1.1 Definition:

Let $t \in \mathcal{S}'(\mathbb{R}^d)$, and denote by $B(t)$ the following set:

$$B(t) = \{a \in \mathbb{R}'^d; e^{-(a,x)} t \in \mathcal{S}'(\mathbb{R}^d)\}.$$

First we show:

II.1.2 Lemma:

Let $t \in \mathcal{S}'(\mathbb{R}^d)$; then the set $B(t)$ is a convex subset of $\mathbb{R}'^d$ containing zero.

Proof: Assume $t_1 = e^{-(a_1,x)} t \in \mathcal{S}'(\mathbb{R}^d)$ and $t_2 = e^{-(a_2,x)} t \in \mathcal{S}'(\mathbb{R}^d)$. We now write $t_2 = e^{(a_1-a_2,x)} t_1$. Since $t_1 \in \mathcal{S}'(\mathbb{R}^d)$ one can decompose it into a sum

$$t_1 = t^+ + t^-, \quad t^+, t^- \in \mathcal{S}'(\mathbb{R}^d)$$

with

$$\operatorname{supp} t^+ \subset \{x; (a_1 - a_2, x) \geq 0\},$$
$$\operatorname{supp} t^- \subset \{x; (a_1 - a_2, x) \leq 0\}.$$

From $t_2 = e^{(a_1-a_2,x)} t_1 \in \mathcal{S}'(\mathbb{R}^d)$ one sees that $t^{\pm} e^{(a_1-a_2,x)} \in \mathcal{S}'(\mathbb{R}^d)$. Since for $\lambda \geq 0$ the expression $e^{-\lambda(a_1-a_2,x)}$ and all its derivatives are bounded on $(a_1 - a_2, x) \geq 0$, it follows that $e^{-\lambda(a_1-a_2,x)} t^+ e^{(a_1-a_2,x)} = e^{(1-\lambda)(a_1-a_2,x)} t^+ \in \mathcal{S}'(\mathbb{R}^d)$. By the same argument one finds $e^{(1-\lambda)(a_1-a_2,x)} t^- \in \mathcal{S}'(\mathbb{R}^d)$ and hence $e^{(1-\lambda)(a_1-a_2,x)} t_1 = e^{-(\lambda a_1 + (1-\lambda) a_2, x)} t \in \mathcal{S}'(\mathbb{R}^d)$. $\square$

As a consequence of this Lemma one obtains the following result for the Fourier transform of the tempered distribution t:

II.1.3 Theorem:

Let $t \in \mathcal{S}'(\mathbb{R}^d)$ such that $\overset{\circ}{B}(t)$ is not empty. Denote by $T(\overset{\circ}{B}(t))$ the tube

$$T(\overset{\circ}{B}(t)) = \{z \in \mathbb{C}^d; \operatorname{Im} z \in \overset{\circ}{B}(t)\}.$$

Then, for $q \in \overset{\circ}{B}(t)$, $\mathcal{F}(e^{-(q,x)} t)(p)$ is a function $F(p + iq)$ of the complex variable $z = p + iq$ which is holomorphic in the tube $T(\overset{\circ}{B}(t))$, and $\mathcal{F}(t)$ is the boundary value of $F(p + iq)$ in the sense of tempered distributions,

$$(\mathcal{F}t)(p) = \lim_{\substack{q \in \overset{\circ}{B}(t) \\ q \to 0}} F(p + iq).$$

Proof: Let $q_0 \in \overset{\circ}{B}(t)$. Then there exists a ball of radius r around q_0 which also belongs to $\overset{\circ}{B}(t)$. But this shows that $e^{-(q_0,x)}t$ is a distribution decreasing like $e^{-r\|x\|}$ in every direction. Hence $e^{-(q_0,x)}t$ can be applied to the functions $e^{i(p+iq,x)}$ provided $\|q\| < \frac{r}{2}$. Now in this domain these test-functions depend differentiably on p and q and one obtains

$$\frac{\partial}{\partial p + iq}\left(te^{-i(q_0,x)}, e^{i(p+iq,x)}\right) = \left(te^{-i(q_0,x)}, \frac{\partial}{\partial p + iq}e^{i(p+iq,x)}\right) = 0.$$

But this shows that

$$(t, e^{i(p+iq,x)}) = (te^{-(q,x)}, e^{i(p,x)}) = \mathcal{F}(te^{-(q,x)})$$

is an analytic function in $T(\overset{\circ}{B}(t))$.

Since $(t, e^{i(p+iq,x)})$ is an analytic function in $T(\overset{\circ}{B}(t))$ it follows that the second statement is trivial if zero belongs to $\overset{\circ}{B}(t)$. If zero belongs to the boundary of $B(t)$ then we decompose t again into two pieces

$$t = t^+ + t^-$$

with supp $t^+ \subset \{x; (q,x) \geq 0\}$, and supp $t^- \subset \{x; (q,x) \leq 0\}$, where q is a point belonging to $\overset{\circ}{B}(t)$. Since q and 0 both belong to $B(t)$ one has $t^+ \in S'(\mathbb{R}^d)$ and $t^-e^{-(q,x)} \in S'(\mathbb{R}^d)$.

Now let f be a test function in $S(\mathbb{R}^d)$. Then $e^{-\lambda(q,x)}f$ converges to f on $\{x; (q,x) \geq 0\}$ and $e^{(1-\lambda)(q,x)}f$ converges to $e^{(q,x)}f$ on $\{x; (q,x) \leq 0\}$ for $\lambda \to 0$. Hence for given $\epsilon > 0$ one can find λ^+ and λ^- both larger than zero with

$$|(t^+(e^{-\lambda(q,x)} - 1), f)| \leq \frac{\epsilon}{2}, \quad \text{for} \quad 0 \leq \lambda \leq \lambda^+,$$

and

$$|(t^-e^{-(q,x)}(e^{(1-\lambda)(q,x)} - e^{(q,x)}), f)| \leq \frac{\epsilon}{2}, \quad \text{for} \quad 0 \leq \lambda \leq \lambda^-.$$

But this implies:

$$|(t(e^{-\lambda(q,x)} - 1), f)| \leq \epsilon \quad \text{for} \quad 0 \leq \lambda \leq \min(\lambda^+, \lambda^-),$$

and hence $te^{-\lambda(q,x)}$ converges weakly to t and consequently also strongly since $t \in S'(\mathbb{R}^d)$. Since the Fourier transformation is a continuous operation on $S'(\mathbb{R}^d)$ one sees also that, as λ tends to zero, $\mathcal{F}(te^{-\lambda(q,x)})$ converges to $\mathcal{F}(t)$. $\square$

It might happen that $B(t)$ is contained in a linear subspace of $\mathbb{R}^d$, say in $\mathbb{R}^e$ with $e < d$, and that $B(t)$ has interior points as a set of $\mathbb{R}^e$. In order to describe this situation one uses the concept of distributions with values in some other locally convex topological vector space $\mathcal{V}$. We say $t \in \mathcal{S}'(\mathbb{R}^e; \mathcal{V})$ if for every $f \in \mathcal{S}(\mathbb{R}^e)$ the expression (t, f) is defined and belongs to $\mathcal{V}$ and moreover if the map $t : \mathcal{S} \longrightarrow \mathcal{V}$ is continuous. For our situation we write $\mathbb{R}^d = \mathbb{R}^e \oplus \mathbb{R}^f$ with $e + f = d$ and

$$\mathcal{S}'(\mathbb{R}^d) = \mathcal{S}'(\mathbb{R}^e; \mathcal{S}'(\mathbb{R}^f)).$$

Using this isomorphism one sees that one can define the partial Fourier transformation with respect to the space $\mathbb{R}^e$, which we will denote by $\mathcal{F}_{\mathbb{R}^e}(.)$. Using these concepts the last theorem can easily be generalized.

II.1.4 Corollary:

Suppose $t \in \mathcal{S}'(\mathbb{R}^d)$ such that $B(t)$ is contained in some linear subspace $\mathbb{R}^e$ of $\mathbb{R}^d$ and assume that $\overset{\circ}{B}(t)$ is not empty as a subset of $\mathbb{R}^e$. Then the partial Fourier transformation

$$\mathcal{F}_{\mathbb{R}^e}\left(te^{-(qx)}\right)(p), \qquad p \in \mathbb{R}^e$$

is, for $q \in \overset{\circ}{B}(t)$, a function $F(p + iq)$ of the complex variable $z = p + iq$ with values in $\mathcal{S}'(\mathbb{R}^f)$. $\mathbb{R}^f$ is a complementary subspace of $\mathbb{R}^e$ i.e. $\mathbb{R}^e \oplus \mathbb{R}^f = \mathbb{R}^d$. Moreover, the function $F(z)$ is holomorphic in the tube $T(\overset{\circ}{B}(t)) \subset \mathbb{C}^e$, and $\mathcal{F}_{\mathbb{R}^e}(t)$ is the boundary value of $F(p + iq)$ in the sense of the space $\mathcal{S}'(\mathbb{R}^e; \mathcal{S}'(\mathbb{R}^f))$:

$$\mathcal{F}_{\mathbb{R}^e}(t) = \lim_{\substack{q \in \overset{\circ}{B}(t) \\ q \to 0}} F(p + iq).$$

The proof of the Corollary is a straightforward transcription of the proof of Theorem II.1.3 to the present situation and will, therefore, be omitted.

We have seen that under suitable conditions the Fourier transform of a tempered distribution has an analytic continuation into the tube $T(\overset{\circ}{B}(t))$ and that $\mathcal{F}(t)$ itself is the boundary value of this analytic function. Next we ask for conditions under which an analytic function holomorphic in $T(\overset{\circ}{B}(t))$, with $0 \in \partial B(t)$, has a boundary value in the sense of distributions. The answer is given in the following

II.1.5 Theorem:

Let $\overset{\circ}{B}$ be a convex open set in $\mathbb{R}^d$ and let B be its closure. Assume the properties:

(α) $0 \in \partial B$.

(β) There exists a closed convex cone $\overline{C}$ with interior points such that
$$\overline{C}_1 = \{y \in \overline{C}; \|y\| \le 1\} \subset \overset{\circ}{B} \cup \{0\}.$$

Let $F(x + iy)$ be a function holomorphic in $T(\overset{\circ}{B})$ such that for every fixed $y \in \overset{\circ}{B}$ we have $F(x + iy) \in \mathcal{S}'$; then

$$\lim_{\substack{y \in C_1 \\ y \to 0}} F(x + iy)$$

exists in the sense of tempered distributions if and only if there exists $c > 0$, and $n, m \in \mathbb{N}$ such that F satisfies the estimate

$$|F(x + iy)| \leq c(1 + \|x\|)^n \|y\|^{-m}, \quad \text{for} \quad y \in \overline{C}_1 \setminus \{0\}.$$

Proof: Let y_0 be a fixed point in $\overline{C}_1 \setminus \{0\}$. One wants to study $F(x + i\lambda y_0)$. To this end write $\mathbb{C}^d = \mathbb{C}^1 \oplus \mathbb{C}^{d-1}$ so that one can set $F(x + i\lambda y_0) = G(x', w)$ with $x' \in \mathbb{R}^{d-1}, w \in \mathbb{C}^1$ and $\text{Im } w = \lambda \|y_0\|$. The estimate for F gives for G

$$|G(x', w)| \leq c(1 + \sqrt{\|x'\|^2 + (\text{Re } w)^2})^n |\text{Im } w|^{-m}$$
$$\leq c(1 + \|x'\| + |\text{Re } w|)^n |\text{Im } w|^{-m}.$$

Now set $G^{(0)}(x', w) = G(x', w)$ and define inductively:

$$G^{(i+1)}(x', w) = \int_{w_0}^{w} G^{(i)}(x', w') dw'.$$

Then one obtains for $i \leq m - 1$ the estimate

$$|G^{(i)}(x', w)| \leq c_i'(1 + \|x'\| + |\text{Re } w|)^{n+i} |\text{Im } w|^{-m+i},$$

and

$$|G^{(m+1)}(x', w)| \leq c_{m+1}'(1 + \|x'\| + |\text{Re } w|)^{n+m+1}.$$

Moreover, one has the relation

$$G(x', w) = \frac{d^i}{dw^i} G^{(i)}(x', w).$$

Using this, one obtains with $w = a + ib$ and $f(x) = g(x', a)$:

$$\int \{F(x + i\lambda_1 y_0) - F(x + i\lambda_2 y_0)\} f(x) dx$$

$$= \int \{G(x', a + ib_1) - G(x', a + ib_2)\} g(x', a) dx' da$$

$$= (-1)^m \int \{G^m(x', a + ib_1) - G^m(x', a + ib_2)\} \frac{d^m}{da^m} g(x', a) dx' da$$

$$= (-1)^m \int \left(\int_{b_2}^{b_1} \frac{d}{dc} G^{m+1}(x', a + ic) dc \right) \frac{d^m}{da^m} g(x', a) dx' da$$

$$= (-1)^{m+1} \int \left(\int_{b_2}^{b_1} G^{m+1}(x', a + ic) dc \right) \frac{d^{m+1}}{da^{m+1}} g(x', a) dx' da.$$

From this one obtains with the above estimate

$$\left| \int \{ F(x + i\lambda_1 y_0) - F(x + i\lambda_2 y_0) \} f(x) dx \right|$$

$$\leq |b_1 - b_2| c'_{m+1} \int (1 + \|x'\| + |a|)^{n+m+1} \left| \frac{d^{m+1}}{da^{m+1}} g(x', a) \right| dx' da,$$

which shows that $F(x + i\lambda y_0)$ is a Cauchy sequence in $\mathcal{S}'$. Now this formula is easily generalized to d variables and gives:

$$\left| \int \{ F(x + iy_1) - F(x + iy_2) f(x) dx \right|$$

$$\leq c'' \|y_1 - y_2\| \int (1 + \|x\|_1)^{n+m+1} \max_{|i| \leq m+1} |D^i f(x)| dx$$

where $\|x\|_1 = \sum_i |x_i|$ is a norm on $\mathbb{R}^d$. But this formula shows that the limit exists for y in $\overline{C}_1$.

Conversely assume that $F(x + iy)$ has a limit in $\mathcal{S}'$. Denote this limit by $\mathcal{F}(t)$, i.e. t is the inverse Fourier transform of the limit. Then we obtain:

$$F(x + iy) = \mathcal{F}(te^{-(y,p)}), \qquad y \in \overline{C}_1.$$

Now one can write

$$t = P(D_p)Q(p)\widehat{t}$$

where P and Q are suitable polynomials and $\widehat{t} \in \mathcal{L}^1$, so that

$$\mathcal{F}(t) = P(-ix)Q(iD_x)\mathcal{F}(\widehat{t})$$

and

$$F(x + iy) = P(-iz)Q(iD_z)\widehat{F}(z)$$

where $\widehat{F}(z)$ is bounded in $T(C_1)$. Estimating the derivatives of $\widehat{F}$ by Cauchy's integral formula we obtain the estimate given in the theorem. $\quad\square$

Knowing under what conditions boundary values of analytic functions exist in $\mathcal{S}'$ one can now look at the relations between support properties of tempered distributions and analytic properties of their Fourier transforms.

II.1.6 Proposition:

Let $t \in \mathcal{S}'(\mathbb{R}^d)$ and let $x_0 \in \mathbb{R}'^d$ with $\|x_0\| = 1$. Then the following statements are equivalent:

(1) $\operatorname{supp} t \subset \{ p; (p, x_0) \geq a \}, \qquad a \in \mathbb{R}.$
(2) Writing $\mathbb{R}^d = \mathbb{R}^1 \oplus \mathbb{R}^{d-1}$ where $\mathbb{R}^{d-1}$ is the subspace $(p, x_0) = 0$, the partial Fourier transform $\mathcal{F}_{\mathbb{R}^1}(t)$ has an analytic continuation into the upper half plane $\operatorname{Im} z > 0$, and there exists for every seminorm N on $\mathcal{S}'(\mathbb{R}^{d-1})$ a constant c with

$$N\big(\mathcal{F}_{\mathbb{R}^1}(t)(x + iy)\big) \leq c(1 + |x|)^n (1 + |y|^{-1})^m e^{-a|y|}.$$

Proof: Assume (1); then $te^{-\lambda(p,x)} \in \mathcal{S}'$ for $\lambda > 0$. Hence by Corollary II.1.4 the partial Fourier transform has an analytic continuation into the upper half plane. Moreover, on the set $\{p; (p, x_0) \geq a\}$ one has the estimate $e^{-\lambda(p,x_0)} \leq e^{-ay}$ for $\lambda = y > 0$ so that one obtains for $f \in \mathcal{S}(\mathbb{R}^{d-1})$

$$\mathcal{F}_{\mathbb{R}^1}(t, fe^{-(p,yx_0)}) = e^{-ay}\mathcal{F}_{\mathbb{R}^1}(t, fe^{ay-(p,yx_0)}).$$

Since $e^{ay-(p,yx_0)}$ is bounded on the support of t one obtains by Theorem II.1.5

$$|\mathcal{F}_{\mathbb{R}^1}(t, fe^{-(p,yx_0)})| \leq c(f)(1 + |x|)^n(1 + |y|^{-1})^m e^{-ay}.$$

From $t \in \mathcal{S}'(\mathbb{R}^d)$ it follows that (t, f) is a bounded set of distributions when f is taken from a bounded set $Bd \subset \mathcal{S}(\mathbb{R}^{d-1})$. Hence the constant $c(f)$ is bounded for $f \in Bd$, which is equivalent to the statement (2).

Next, assume (2). Then for $y > 0$ a function $\frac{1}{(i+z)^{n+2}}\mathcal{F}_{\mathbb{R}^1}(t)(z)$ belongs to $\mathcal{L}^1$ with values in $\mathcal{S}'(\mathbb{R}^{d-1})$. Since it is bounded by e^{-ay} for large y it follows that $\mathcal{F}_{\mathbb{R}^1}^{-1}\left(\frac{1}{(i+z)^{n+2}}\mathcal{F}_{\mathbb{R}^1}(t)\right) = \tilde{t}$ has its support in $\{p; (p, x_0) \geq a\}$. But since $t = (i + (x_0, iD))^{n+2}\tilde{t}$ one sees that the support of t is the same as that of $\tilde{t}$. □

This last result leads us to a characterization of the Fourier transform of a tempered distribution which has support in a cone.

II.1.7 Theorem:

Let $\overline{C} \subset \mathbb{R}^d$ be a closed, convex, proper cone with interior points. Denote by C' its dual cone in $\mathbb{R}'^d$ i.e. $C' = \{x; (x, p) \geq 0, \quad \text{for} \quad p \in \overline{C}\}$. Let $\overset{\circ}{C'}$ be the interior of C' and for $y \in \overset{\circ}{C'}$ denote by $d(y)$ the distance of y from the boundary of C'. Then the following statements are equivalent:

(1) *$t \in \mathcal{S}'$ has support in $a + \overline{C}$.*

(2) *$\mathcal{F}(t)$ is the boundary value of an analytic function holomorphic in the tube $T(\overset{\circ}{C'})$ and there exist natural numbers N and M and $k > 0$ with*

$$|e^{(ay)}\mathcal{F}(t)(x + iy)| \leq k(1 + \|x\|)^N(1 + (d(y))^{-1})^M, \qquad y \in \overset{\circ}{C'}.$$

Proof: Assume (1) which implies that $B(t)$ contains the cone C'. Then by Theorem II.1.3 $\mathcal{F}(t)$ is holomorphic in the tube $T(\overset{\circ}{C'})$. Since t is tempered there exists an $n \in \mathbb{N}$ such that $t(1 + \|p\|^2)^{-n}$ is a finite derivative of an $\mathcal{L}^1$-function. Hence $\mathcal{F}(t(1 + \|p\|^2)^{-n})$ is a polynomially bounded continuous function, which means that

$$|e^{(a,y)}\mathcal{F}(t(1 + \|p\|^2)^{-n})(x + iy)| \leq k(1 + \|x\|)^N, \qquad y \in C'.$$

Since $\mathcal{F}(t)(x+iy) = P(iD)\mathcal{F}\big(t(1+\|p\|^2)^{-n}\big)(x+iy)$, we obtain the above estimate by using the Cauchy formula for the derivative. Conversely, if this estimate is fulfilled then we have by Proposition II.1.6 that supp $t \subset \{p; (p,b) \geq (a,b)\}$ with $b \in \overset{\circ}{C'}$. Since this holds for every $b \in \overset{\circ}{C'}$ we obtain by taking the intersection of these half spaces:

$$\operatorname{supp} t \subset a + \overline{C}. \hspace{3cm} \square$$

II.2   Symmetry groups and continuity

Let $\mathcal{A}$ be a C^*-algebra and $G(\tau)$ be a topological group. Assume G acts as group of automorphisms on $\mathcal{A}$, which means that one has a map $\alpha : G \to \operatorname{Aut}\mathcal{A}$. No continuity is required for the map α with respect to τ. Such a system is called a C^*-dynamical system and denoted by the triple $\{\mathcal{A}, G, \alpha\}$.

II.2.1 Definition:

By $\mathcal{A}_c^*$ we denote the set of $\varphi \in \mathcal{A}^*$ ($\mathcal{A}^*$ denotes the topological dual of $\mathcal{A}$), such that for every $\epsilon > 0$ exists a neighbourhood $\mathcal{U}$ of the identity of G such that

$$\|\varphi \circ \alpha_g - \varphi\| \leq \epsilon \quad \text{for} \quad g \in \mathcal{U}.$$

It is the aim of this section to prove the following

II.2.2 Theorem:

Let $\{\mathcal{A}, G, \alpha\}$ be a C^*-dynamical system and assume $G(\tau)$ is a topological group. Then the space $\mathcal{A}_c^*$ has the following properties:

(i) $\mathcal{A}_c^*$ is a linear space.

(ii) $\mathcal{A}_c^*$ is closed in the norm-topology.

(iii) $\mathcal{A}_c^*$ is invariant under the action of the group, i.e. $\varphi \in \mathcal{A}_c^*$ implies $\varphi \circ \alpha_g \in \mathcal{A}_c^*$ for every $g \in G$.

(iv)   With $\varphi \in \mathcal{A}_c^*$ one finds also that φ^* and $|\varphi|$ belong to $\mathcal{A}_c^*$.

(v) $\mathcal{A}_c^*$ is generated by its positive elements.

(vi)   Let β be an automorphism of $\mathcal{A}$ commuting with every α_g, $g \in G$ then $\beta \mathcal{A}_c^* = \mathcal{A}_c^*$.

(vii)   Let $\mathcal{A}_G^{**}$ be the sub-von-Neumann algebra of $\mathcal{A}^{**}$ of elements which are α_g-invariant for all $g \in G$, then $\varphi \in \mathcal{A}_c^*$ and $y \in \mathcal{A}_G^{**}$ implies $y\varphi$ and $\varphi y \in \mathcal{A}_c^*$.

(We have used the notations $(\varphi y)(x) = \varphi(xy)$ and $(y\varphi)(x) = \varphi(yx)$.)

Proof: (trivial part) (i) Given $\epsilon > 0$ and $\varphi_1, \varphi_2 \in \mathcal{A}_c^*$, there exist two neighbourhoods $\mathcal{U}_1(\epsilon)$ and $\mathcal{U}_2(\epsilon)$. Define

$$\mathcal{U} = \mathcal{U}_1\big(\frac{\epsilon}{2|\lambda_1|}\big) \cap \mathcal{U}_2\big(\frac{\epsilon}{2|\lambda_2|}\big).$$

Then one obtains

$$\|(\lambda_1\varphi_1 + \lambda_2\varphi_2) \circ \alpha_g - (\lambda_1\varphi_1 + \lambda_2\varphi_2)\| \leq$$
$$|\lambda_1| \|\varphi_1 \circ \alpha_1 - \varphi_1\| + |\lambda_1| \|\varphi_2 \circ \alpha_2\| \leq$$
$$|\lambda_1|\frac{\epsilon}{2|\lambda_1|} + |\lambda_2|\frac{\epsilon}{2|\lambda_1|} = \epsilon$$

provided g belongs to $\mathcal{U}$.

(ii) Assume $\varphi_n \in \mathcal{A}_c^*$ and $\{\varphi_n\}$ converge to φ in norm. For $\epsilon > 0$ there exists n_0 such that $\|\varphi - \varphi_n\| \leq \frac{\epsilon}{3}$ for $n > n_0$. Moreover, there exists for a fixed $n_1 > n_0$ a neighbourhood $\mathcal{U}$ such that $\|\varphi_{n_1} \circ \alpha - \varphi_{n_1}\| \leq \frac{\epsilon}{3}$ for $g \in \mathcal{U}$. Let now $g \in \mathcal{U}$; then one obtains

$$\|\varphi \circ \alpha_g - \varphi\| \leq \|\varphi \circ \alpha_g - \varphi_{n_1} \circ \alpha_g\| + \|\varphi_{n_1} \circ \alpha_g - \varphi_{n_1}\|$$
$$+ \|\varphi_{n_1} - \varphi\| \leq \frac{\epsilon}{3} + \frac{\epsilon}{3} + \frac{\epsilon}{3} = \epsilon.$$

This shows that $\varphi \in \mathcal{A}_c^*$ and hence $\mathcal{A}_c^*$ is closed.

(iii) Let h be a fixed element in G. Then

$$\|\varphi \circ \alpha_h \circ \alpha_g - \varphi \circ \alpha_h\| = \|\varphi \circ \alpha_{hgh-1} - \varphi\|.$$

As $h\mathcal{U}h^{-1}$ is an open neighbourhood of the identity one sees that $\varphi \circ \alpha_h$ belongs to $\mathcal{A}_c^*$ whenever φ belongs to $\mathcal{A}_c^*$.

(iv) Since $\|\varphi^* \circ \alpha_g - \varphi^*\| = \|\varphi \circ \alpha_g - \varphi\|$ the first part of (iv) follows. The second part will be proved separately.

(v) Assume (iv) is correct. Then because of $\varphi = \frac{1}{2}(\varphi + \varphi^*) + \frac{1}{2i}(\varphi - \varphi^*)$ it is sufficient to prove that real φ's are generated by positive ones. If φ is real then we have $\varphi = \frac{1}{2}(|\varphi| + \varphi) - \frac{1}{2}(|\varphi| - \varphi)$. Since by (iv) $|\varphi|$ belongs to $\mathcal{A}_c^*$ and $|\varphi| + \varphi$ and $|\varphi| - \varphi$ are both positive (v) is an implication of (i) and (iv).

(vi) Compute $|(\varphi \circ \beta)(x) - \varphi \circ \beta(\alpha_g x)| = |\varphi(\beta x) - \varphi \circ \alpha_g(\beta x)| \leq \|\varphi - \varphi \circ \alpha_g\| \|\beta x\| = \|\varphi - \varphi \circ \alpha_g\| \|x\|$. Hence $\varphi \in_c^*$ implies $\beta\varphi \in \mathcal{A}_c^*$.

(vii) For $y \in \mathcal{A}_G^{**}$ and $\varphi \in \mathcal{A}_c^*$ one obtains $|\varphi y(x) - \varphi y(\alpha_g x)| = |\varphi(xy) - \varphi(\alpha_g xy)| \leq \|\varphi - \varphi \circ \alpha_g\| \|x\| \|y\|$. Hence one obtains $\|\varphi y - \varphi y \circ \alpha_g\| \leq \|y\| \|\varphi - \varphi \circ \alpha_g\|$ and consequently $\varphi y \in \mathcal{A}_c^*$ for $\varphi \in \mathcal{A}_c^*$ and $y \in \mathcal{A}_G^{**}$. The second part follows from the relation $(\varphi y)^* = y^*\varphi^*$ and the invariance of $\mathcal{A}_c^*$ and $\mathcal{A}_G^{**}$ under involution.

The second statement of (iv) can easily be reduced to the following

II.2.3 Theorem:

Let $\mathcal{A}$ be a C^*-algebra and φ and $\psi \in \mathcal{A}^*$. Then one obtains the estimate

$$\| |\varphi| - |\psi| \| \leq \|\varphi - \psi\| + (\sqrt{\|\varphi\|} + \sqrt{\|\psi\|})\sqrt{\|\varphi - \psi\|}.$$

Proof: Let $\varphi = U|\varphi|$, i.e. $\varphi(x) = |\varphi|(Ux)$, $|\varphi|(x) = \varphi(U^*x)$. Then one has $\|\varphi\| = \| \, |\varphi| \, \| = |\varphi|(1) = \varphi(U^*)$ and similar formulas for $\psi = V|\psi|$. With these notations one obtains

$$\| \, |\varphi| - |\psi| \, \| = \sup_{\|x\|=1} (|\varphi|(x) - |\psi|(x))$$

$$\leq \sup_{\|x\|=1} ((|\varphi|(x) - \varphi(V^*x)) + (\varphi(V^*x) - \psi(V^*x)))$$

$$\leq \sup_{\|x\|=1} (|\varphi|(x - UV^*x)) + \|\varphi - \psi\|$$

$$\leq \sup_{\|x\|=1} [|\varphi|(x^*x)]^{1/2} [|\varphi|(\{1 - UV^*\}\{1 - VU^*\})]^{1/2} + \|\varphi - \psi\|$$

$$\leq \|\varphi\|^{1/2} [|\varphi|(\{1 - UV^*\}\{1 - VU^*\})]^{1/2} + \|\varphi - \psi\|.$$

Using

$$|\varphi|((1 - UV^*)(1 - VU^*)) = |\varphi|(1) + |\varphi|(UV^*VU^*) - 2\operatorname{Re}|\varphi|(UV^*)$$

$$\leq 2(\|\varphi\| - \operatorname{Re}|\varphi|(UV^*)) = 2(\|\varphi\| - \varphi(V^*))$$

$$\leq 2|\, \|\varphi\| - \psi(V^*)| + 2|\psi(V^*) - \varphi(V^*)|$$

$$\leq 2|\, \|\varphi\| - \|\psi\| \, | + 2\|\varphi - \psi\|$$

$$\leq 4\|\varphi - \psi\|$$

we obtain

$$\| \, |\varphi| - |\psi| \, \| \leq \|\varphi - \psi\| + 2\sqrt{\|\varphi\|}\sqrt{\|\varphi - \psi\|}.$$

Interchanging the role of φ and ψ we find

$$\| \, |\varphi| - |\psi| \, \| \leq \|\varphi - \psi\| + 2\sqrt{\|\psi\|}\sqrt{\|\varphi - \psi\|}.$$

Taking the average we get the final result:

$$\| \, |\varphi| - |\psi| \, \| \leq \|\varphi - \psi\| + (\sqrt{\|\varphi\|} + \sqrt{\|\psi\|})\sqrt{\|\varphi - \psi\|}. \qquad \square$$

With this theorem we also have proved statement (iv) of theorem II.2.2.

II.3 Spaces of momentum transfer

In this section the group in question will be the translation group $\mathbb{R}^d$. Let $\{\mathcal{A}, \mathbb{R}^d, \alpha\}$ be a C^*–dynamical system and let $\mathcal{A}_c^*$ be the subspace of linear functionals such that $\mathbb{R}^d \ni a \to \varphi \circ \alpha_a$ is strongly continuous. This space is defined in II.2.1 and its properties are described in Theorem II.2.2. If $x \in \mathcal{A}^{**}$ and $\varphi \in \mathcal{A}_c^*$ then $\varphi(\alpha_a x)$ is a continuous bounded function on $\mathbb{R}^d$. Hence for any $f \in \mathcal{L}^1(\mathbb{R}^d)$ the expression $\int \varphi(\alpha_a x) f(a) da$ is well–defined and represents a continuous linear functional on $\mathcal{A}_c^*$.

II.3.1 Definition:

(α) *For* $f \in \mathcal{L}^1(\mathbb{R}^d)$ *define*

$$[x(f)] = \{y \in \mathcal{A}^{**}; \quad \text{for every } \varphi \in \mathcal{A}^*_c \text{ one has}$$

$$\varphi(y) = \int \varphi(\alpha_a x) f(a) \, \mathrm{d}a\}.$$

This definition makes sense since $\mathcal{A}^*_c$ *is a linear space.*

(β) *By* N_c *we denote the subspace of elements in* $\mathcal{A}^{**}$ *which annihilate* $\mathcal{A}^*_c$. *Since* $\mathcal{A}^*_c$ *is a norm-closed subspace of* $\mathcal{A}^*$ *it follows that* N_c *is ultraweakly closed.*

Next we want to define the spaces of momentum transfer. To this end remark that the dual group of $\mathbb{R}^d$ is again $\mathbb{R}^d$ which will be denoted by $\mathbb{R}'^d$.

II.3.2 Definition:

Let D be a closed set in $\mathbb{R}'^d$. We denote by $M(D)$ the following set of operators:

$$M(D) = \{x \in \mathcal{A}^{**}; [x(f)] \subset N_c \quad \text{for every} \quad f \in \mathcal{L}^1(\mathbb{R}^d)$$

$$\text{with} \quad \operatorname{supp} \mathcal{F}^{-1}(f) \subset (\mathbb{R}'^d) \setminus D\}.$$

Equivalently for every $\varphi \in \mathcal{A}^*_c$ *one has* $\operatorname{supp} \mathcal{F}^{-1}\varphi(\alpha_a x) \subset D$ *(where the Fourier transform is taken in the sense of tempered distributions). If D is the empty set then we will identify $M(\emptyset)$ with N_c.*

With these notations one obtains the following result:

II.3.3 Proposition:

The spaces $M(D)$ of momentum transfer have the following properties:

(i) $M(D)$ *is a linear space which is* σ*-weakly closed.*

(ii) $M(-D) = M(D)^*$.

(iii) $D_1 \subset D_2$ *implies* $M(D_1) \subset M(D_2)$.

(iv) $M(\cap_\beta D_\beta) = \cap_\beta M(D_\beta)$.

(v) *Assume $D_1 \cap D_2 = \emptyset$ and one of the D_i is compact. Then*

$$M(D_1 \cup D_2) = M(D_1) + M(D_2).$$

(vi) $\alpha_a M(D) = M(D)$ *for every* $a \in \mathbb{R}^d$.

(vii) *Let $\mathcal{A}^{**}_G$ be the set of elements in $\mathcal{A}^{**}$ which are pointwise invariant under the action of the group $\mathbb{R}^d$. Then:*

$$\mathcal{A}^{**}_G \subset M(\{0\}).$$

(viii) $M(\mathbb{R}'^d) = \mathcal{A}^{**}$.

(ix) *Let β be an automorphism commuting with the translations α_g. Then*

$$\beta M(D) = M(D).$$

(x) *For* $y \in \mathcal{A}_G^{**}$,

$$M(D)y \subset M(D) \quad \text{and} \quad yM(D) \subset M(D).$$

Proof: (i) $x \in M(D)$ means for every $f \in \mathcal{L}^1(\mathbb{R}^d)$ with supp $\mathcal{F}^{-1}f \cap D = \emptyset$ and every $\varphi \in \mathcal{A}_c^*$ one has $\varphi_f(x) = 0$ with

$$\varphi_f := \int \varphi \circ \alpha_a f(a) \, da.$$

Therefore

$$M(D) = \cap \{\text{kernel } \varphi_f; \varphi \in \mathcal{A}_c^* \quad \text{and supp } \mathcal{F}^{-1}f \cap D = \emptyset \},$$

which is a weakly closed set, and which is a linear space since it is the intersection of linear spaces.

(ii) The relation $x \in M(D)$ is equivalent to the condition that for every $\varphi \in \mathcal{A}_c^*$ one has supp $\mathcal{F}^{-1}\varphi(\alpha_a x) \subset D$. Since with $\varphi \in \mathcal{A}_c^*$ and $\varphi^* \in \mathcal{A}_c^*$ one obtains, by complex conjugation, supp $\mathcal{F}^{-1}\varphi^*(\alpha_a x^*) = \text{supp } \overline{\mathcal{F}^{-1}\varphi(\alpha_a x)}$ $\subset \{-D\}$.

(iii) This follows immediately by the characterization of $M(D)$ used in (ii).

(iv) From (iii) follows $M(\cap_\beta D_\beta) \subset \cap_\beta M(D_\beta)$. On the other hand if $x \in \cap_\beta M(D_\beta)$ then it follows that for every $\varphi \in \mathcal{A}_c^*$ one has supp $\mathcal{F}^{-1}\varphi(\alpha_a x) \subset D_\beta$ for every β and consequently in $\cap_\beta D_\beta$. But this shows $x \in M(\cap_\beta D_\beta)$ by the second definition of $M(D)$.

(v) Assume D_1 is compact. Since D_2 is closed and the two have empty intersection there exists a function $f(p) \in C^\infty(\mathbb{R}'^d)$ with compact support and $f(p) = 1$ for $p \in D_1$, $f(p) = 0$ for $p \in D_2$. Denote the Fourier transform of f by $\tilde{f}$. Now let $x \in M(D_1 \cup D_2)$. Take $x_1 \in [x(\tilde{f})]$ and define $x_2 = x - x_1$. Then we have $x = x_1 + x_2$. Next let $g \in \mathcal{L}^1(\mathbb{R}^d)$ with support $\mathcal{F}^{-1}g \cap D_1 = \emptyset$. Then for $\varphi \in \mathcal{A}_c^*$ it follows that:

$$\int \varphi(\alpha_a x_1)g(a) \, da = \int \varphi(\alpha_{a+b}x_1)\tilde{f}(b)g(a) \, db \, da$$

$$= \int \varphi(\alpha_c x)\tilde{f}(c-a)g(a) \, dc \, da = 0.$$

Since f vanishes in D_2 and $\mathcal{F}^{-1}g$ in D_1 it follows that $\mathcal{F}^{-1}(\tilde{f} \star g)$ vanishes in $D_1 \cup D_2$.

This implies $x \in M(D_1)$. Since $(1 - f(p))$ vanishes in D_1 it follows by the same argument that $x \in M(D_2)$.

(vi) This follows from the fact that the Fourier transformations of $f(a)$ and $f_b(a) = f(a - b)$ have the same support.

(vii) If $\alpha_a x = x$ is independent of a then the inverse Fourier transformation is proportional to a δ-function at the origin and hence $x \in M(\{0\})$.

(viii) This follows directly from the definition of $M(D)$.

(ix) From Theorem II.2.2 (vi) we know that $\mathcal{A}_c^*$ is invariant under β. But this implies the statement by the definition of $M(D)$.

(x) This follows from the same arguments as the statement before, since $\mathcal{A}_c^*$ is invariant under multiplication by elements from $\mathcal{A}_G^{**}$ on the right as well as on the left (Theorem II.2.2.(vi)). □

For later use we shall need the following result:

II.3.4 Lemma:

Denote by $D_r \subset \mathbb{R}^{\prime d}$ the ball of radius r centered at zero.

(i) $x \in \mathcal{A}^{**}$ belongs to $M(D_r)$ if and only if for every $\varphi \in \mathcal{A}_c^*$ the function

$$a \to \varphi(\alpha_a x)$$

extends to an entire analytic function $W(z)$ and this function obeys the estimate

$$|W(z)| \leq \|\varphi\| \, \|x\| \exp\{r\|\operatorname{Im} z\|\}.$$

(ii) For the above x there exist $x_i \in M(D_r)$ such that

$$[\alpha_a x] = \sum \frac{a^i}{i!}[x_i].$$

Here i denotes the multi-index $i = (i_1..., i_d)$ with $|i| = \sum i_j$, and $a^i = \prod(a_j)^{i_j}$, and $i! = \prod(i_j)!$.

If x is such that for every $\varphi \in \mathcal{A}_c^*$ one has the estimate $|\varphi(x)| \leq \|\varphi\| n(x)$, then one obtains

$$|\varphi(x_i)| \leq \|\varphi\| n(x) r^{|i|}$$

where $n(x)$ is the semi-norm $\sup\{|\varphi(x)|; \varphi \in \mathcal{A}_c^*, \|\varphi\| = 1\}$.

Proof: (i) By the definition of $M(D_r)$ it follows that $\mathcal{F}^{-1}\varphi(\alpha_a x)$ has its support in D_r for $\varphi \in \mathcal{A}_c^*$. Since $\varphi(\alpha_a x)$ is bounded for real x by $\|\varphi\| \, \|x\|$ one obtains the estimate by a Phragmén–Lindelöf type argument since $W(z)$ is of exponential type r.

(ii) Since $\varphi(\alpha_a x)$ is entire it may be written in the form

$$\varphi(\alpha_a x) = \sum \frac{a^i}{i!} C_i.$$

As $\varphi(\alpha_a x)$ is of exponential type r,

$$|C_i| \leq \sup_{a \in \mathbb{R}^d} |\varphi(\alpha_a x)| r^{|i|} \leq \|\varphi\| n(x) r^{|i|}$$

with $n(x)$ as defined above, which implies $n(x) \leq \|x\|$. This estimate shows that C_i is a continuous linear functional on $\mathcal{A}_c^*$ with norm $n(x) r^{|i|}$. Therefore, by the Hahn–Banach theorem we can find an extension x_i with the same norm such that $\varphi(x_i) = C_i$ for every $\varphi \in \mathcal{A}_c^*$. Finally

$$\varphi(\alpha_a x_i) = \frac{\partial^{|i|}}{\partial_{a_1}^{i_1} \dots \partial_{a_d}^{i_d}} \varphi(\alpha_a x).$$

But these functions are of the same exponential type, which means that $\mathcal{F}^{-1}\varphi(\alpha_a x_i)$ and $\mathcal{F}^{-1}\varphi(\alpha_a x)$ have the same support. This implies $x_i \in M(D_r)$ by definition II.3.2. □

II.4 Spectrum condition: The one-dimensional case

In this section we will investigate the one-parameter group. This means the C^*-dynamical system will be $\{\mathcal{A}, \mathbb{R}, \alpha\}$. The double dual of $\mathcal{A}$ will again be denoted by $\mathcal{A}^{**}$ and $\mathcal{A}_G^{**}$ will be the α_a–invariant elements of $\mathcal{A}^{**}$, and $\mathcal{A}_c^*$ the space of functionals such that $a \to \varphi \circ \alpha_a$ is continuous in the norm-topology on $\mathcal{A}^*$. For D a closed subset of $\mathbb{R}'$, the space $M(D)$ is the space of momentum transfer studied in the last section. Now we will introduce the following notation:

II.4.1 Definition:

(i) *For $\lambda \in \mathbb{R}$ let $E(\lambda)$ be the maximal projection in $\mathcal{A}^{**}$ annihilating $M((-\infty, -\lambda])$ from the right, i.e. $E(\lambda)$ is the projection onto the common null space of all $x \in M((-\infty, -\lambda])$.*

(ii) *E^+ will denote the projection*

$$E^+ = \lim_{\lambda \to \infty} E(\lambda).$$

We have to show that this projection exists.

II.4.2 Proposition:

With the notation of Definition II.4.1 we obtain:

(i) *$E(\lambda)$ is monotone increasing and hence E^+ is the strong limit of the $E(\lambda)$.*

(ii) *$E(\lambda)$ is continuous from the left.*

(iii) *$E(\lambda) = 0$ for $\lambda \leq 0$.*

(iv) *$E(\lambda)$ is invariant, more precisely $E(\lambda) \in \mathcal{Z}(\mathcal{A}_G^{**})$ – the centre of $\mathcal{A}_G^{**}$.*

(v) *Let β be an automorphism of $\mathcal{A}$ which commutes with the translations α_a. Then*

$$\beta E(\lambda) = E(\lambda).$$

Proof: (i) Let $\mu < \lambda \in \mathbb{R}$. Then $(-\infty, -\lambda] \subset (-\infty, -\mu]$. Hence $M((-\infty, -\lambda]) \subset M((-\infty, -\mu])$ by Proposition II.3.3. This implies that $E(\mu)$ is a right annihilator of $M((-\infty, -\lambda])$ and consequently $E(\mu) \leq E(\lambda)$.

(ii) Let $\lambda_i \leq \lambda_{i+1}$ converge to λ. Then $M((-\infty, -\lambda]) = \cap_i M((-\infty, -\lambda_i])$ follows again from II.3.3. However, this implies $E(\lambda) = \text{lub} E(\lambda_i)$, and since λ_i is increasing, it follows that $E(\lambda) = \lim_{i \to \infty} E(\lambda_i)$.

(iii) If $\lambda \leq 0$ then it follows that $(-\infty, -\lambda] \ni \{0\}$. This implies by Proposition II.3.3 (vii) that $\mathbb{1} \in M((-\infty, -\lambda])$ and consequently $E(\lambda) = 0$.

(iv) Since $M((-\infty, -\lambda])$ is invariant under α_g it follows that $E(\lambda)$ is an invariant projector. Let $y \in \mathcal{A}_G^{**}$. Then it follows from Prop. II.3.3.(x) that $M((-\infty, -\lambda])$ is invariant under multiplication from the left by $\mathcal{A}_G^{**}$. Hence for $x \in M((-\infty, -\lambda])$ and $y \in \mathcal{A}_G^{**}$ we find that $yE(\lambda)$ annihilates x and consequently $E(\lambda)yE(\lambda) = yE(\lambda)$. Taking now y selfadjoint then it follows that y commutes with $E(\lambda)$ and hence $E(\lambda) \in \mathcal{Z}(\mathcal{A}_G^{**})$.

(v) From Proposition II.3.3 (ix) we know $\beta M((-\infty, -\lambda]) = M((-\infty, -\lambda])$ which implies the relation $\beta E(\lambda) = E(\lambda)$. □

With $E(\lambda) \in \mathcal{Z}(\mathcal{A}_G^{**})$ it follows that $E^+ = \lim_{\lambda \to \infty}$ also belongs to $\mathcal{Z}(\mathcal{A}_G^{**})$. It is our next aim to show that E^+ also belongs to the center of $\mathcal{A}^{**}$. To this end we need some further notation.

II.4.3 Definition:

With the assumptions of this section define

(i) $\mathcal{A}^*(\mathbb{R}'^+) = \{\varphi \in \mathcal{A}^*; E^+\varphi = \varphi E^+ = \varphi\}$.

(ii) $\mathcal{A}_0^*(\mathbb{R}'^+) = \{\varphi \in \mathcal{A}^*;$ there exist $0 < \lambda < \infty$ with $E(\lambda)\varphi = \varphi E(\lambda) = \varphi\}$.

Using this definition one obtains:

II.4.4 Proposition:

With the assumptions and notations of this section one has

(i) $\mathcal{A}^*(\mathbb{R}'^+)$ *is a norm-closed linear subspace of* $\mathcal{A}^*$ *and* $\varphi \in \mathcal{A}^*(\mathbb{R}'^+)$ *implies* $\varphi^* \in \mathcal{A}^*(\mathbb{R}'^+)$.

(ii) $\mathcal{A}_0^*(\mathbb{R}'^+)$ *is norm-dense in* $\mathcal{A}^*(\mathbb{R}'^+)$.

(iii) $\mathcal{A}_0^*(\mathbb{R}'^+) \subset \mathcal{A}_c^*$.

(iv) *If* $\varphi \in \mathcal{A}_0^*(\mathbb{R}'^+)$ *then* $a \to \alpha_a$ *extends to an entire analytic function and there exists a constant* $\lambda > 0$ *with*

$$\|\varphi \circ \alpha_z\| \leq \|\varphi\| e^{\lambda |\text{Im } z|}.$$

(v) $\varphi \in \mathcal{A}^*(\mathbb{R}'^+)$ *and* $x \in \mathcal{A}^{**}$ *implies that* $x\varphi \in \mathcal{A}_c^*$ *and* $\varphi x \in \mathcal{A}_c^*$.

(vi) $\varphi \in \mathcal{A}_0^*(\mathbb{R}'^+)$ *and* $x \in M([-\mu, \mu])$ *imply that* $x\varphi$ *and* φx *both belong to* $\mathcal{A}_0^*(\mathbb{R}'^+)$.

(vii) *If* $\varphi \in \mathcal{A}^*(\mathbb{R}'^+)$ *and* $x, y \in \mathcal{A}^{**}$ *then* $x\varphi y$ *belongs to* $\mathcal{A}^*(\mathbb{R}'^+)$.

(viii) *The projection* E^+ *belongs to the center of* $\mathcal{A}^{**}$ *which is equivalent to the statement that* $\mathcal{A}^*(\mathbb{R}'^+)$ *is a folium.*

Proof: (i) Let $\varphi_i \in \mathcal{A}^*(\mathbb{R}'^+)$ be a norm convergent sequence with limit φ. Then one has for $x \in \mathcal{A}$ the equation $\varphi_i(E^+ x) = \varphi_i(xE^+) = \varphi_i(x)$, which implies the same relation for the limit φ and hence $\varphi \in \mathcal{A}^*(\mathbb{R}'^+)$. The relation $(\varphi x)^* = x^* \varphi^*$ implies that $\mathcal{A}^*(\mathbb{R}'^+)$ is invariant under involution.

(ii) Let $\varphi \in \mathcal{A}^*(\mathbb{R}'^+)$. Since E^+ is the ultra-strong limit of the $E(\lambda)$ which are increasing, there exists for $\epsilon > 0$ a value λ_0 such that $|\varphi|(E(\lambda)) \geq |\varphi|(E^+) - \frac{\epsilon}{2}$ and $|\varphi^*|(E(\lambda)) \geq |\varphi^*|(E^+) - \frac{\epsilon}{2}$ for $\lambda > \lambda_0$. Hence we get

$$\|\varphi - E(\lambda)\varphi E(\lambda)\| \leq \|\varphi - \varphi E(\lambda)\| + \|\varphi E(\lambda) - E(\lambda)\varphi E(\lambda)\| \leq \epsilon.$$

This we obtain because

$$\begin{aligned}
\|\varphi - \varphi E(\lambda)\| &= \sup\{\varphi(xE^+ - xE(\lambda)); \|x\| \leq 1\} \\
&= \sup\{|\varphi|(xE^+ - xE(\lambda)); \|x\| \leq 1\} \\
&= |\varphi|(E^+ - E(\lambda)) \leq \frac{\epsilon}{2} \quad \text{for} \quad \lambda > \lambda_0
\end{aligned}$$

and

$$\begin{aligned}
\|\varphi E(\lambda) - E(\lambda)\varphi E(\lambda)\| &= \sup\{\varphi((E^+ - E(\lambda))xE(\lambda)); \|x\| \leq 1\} \\
&\leq \sup\{\varphi((E^+ - E(\lambda))x); \|x\| \leq 1\} \\
&= \sup\{|\,|\varphi|((E^+ - E(\lambda))x)|; \|x\| \leq 1\} \\
&= |\varphi|((E^+ - E(\lambda))) \leq \frac{\epsilon}{2} \quad \text{for} \quad \lambda > \lambda_0.
\end{aligned}$$

(iii) If $\varphi \in \mathcal{A}_0^*(\mathbb{R}'^+)$ then there exists $\lambda > 0$ such that $\varphi E(\lambda) = \varphi$ and hence φ annihilates $M((-\infty, -\lambda])$. But this set contains N_c and hence $\varphi \in \mathcal{A}_c^*$ by the bipolar theorem.

(iv) If $\varphi \in \mathcal{A}_0^*(\mathbb{R}'^+)$ then there exists $\lambda > 0$ such that $\varphi = \varphi E(\lambda) = E(\lambda)\varphi$. The first relation tells us that $\varphi(x) = 0$ for $x \in M((-\infty, -\lambda])$. If $f \in \mathcal{L}^1(\mathbb{R})$ and supp $\mathcal{F}^{-1}f \subset (-\infty, -\lambda]$ then it follows that $[x(f)] \in M((-\infty, -\lambda])$ and hence $\varphi(\alpha_a x)$ is the Fourier transform of a distribution with support in $[-\lambda, \infty)$. The equation $E(\lambda)\varphi = \varphi$ is equivalent to $\varphi^* E(\lambda) = \varphi^*$. But this implies that $\mathcal{F}^{-1}(\varphi(\alpha_a x))$ has its support in $(-\infty, \lambda]$. Both statements together imply supp $\mathcal{F}^{-1}(\varphi(\alpha_a x)) \subset [-\lambda, \lambda]$. Since $\varphi(\alpha_a x)$ is bounded for real a by $\|\varphi\| \|x\|$ it follows that $\|\varphi \circ \alpha_z\| \leq \|\varphi\| e^{\lambda |\text{Im } z|}$.

(v) If $\varphi \in \mathcal{A}^*(\mathbb{R}'^+), \epsilon > 0$ then there exists φ_1 with $\|\varphi - \varphi_1\| < \epsilon$ and λ with $\varphi_1 E(\lambda) = \varphi_1$. From $0 = M((-\infty, -\lambda])E(\lambda) = xM((-\infty, -\lambda])E(\lambda)$ it follows that φ_1 annihilates $xM((-\infty, -\lambda])$. This means $x\varphi_1$ annihilates N_c

and hence $x\varphi_1 \in \mathcal{A}_c^*$. Since $\mathcal{A}_c^*$ is norm-closed by Theorem II.2.2 one has $x\varphi \in \mathcal{A}_c^*$. Since $\mathcal{A}^*(\mathbb{R}'^+)$ and $\mathcal{A}_c^*$ are both $*$-invariant one finds $x^*\varphi^* \in \mathcal{A}_c^*$ and hence $\varphi x \in \mathcal{A}_c^*$.

(vi) Assume $\varphi E(\lambda) = E(\lambda)\varphi = \varphi$, $x \in M([-\mu, \mu])$ and $y \in \mathcal{A}^{**}$. Then we may write $\varphi(x\alpha_a y) = \varphi(\alpha_a\{(\alpha_{-a}x)y\})$. Since $(\varphi \circ \alpha_a)y \in \mathcal{A}_c^*$ and $\alpha_a x$ is entire analytic we obtain by Lemma II.3.4

$$\varphi(x\alpha_a y) = \varphi(\alpha_a\{(\alpha_{-a}x)y\}) = \sum_{i=0}^{\infty} \frac{(-a)^i}{i!} \varphi(\alpha_a(x_i y))$$

with $n(x_i) \leq \|x\| \, |\mu|^i$. Since $\varphi(\alpha_a(x_i y))$ is of exponential type λ it follows that $\varphi(x\alpha_a y)$ is of exponential type $\lambda + \mu$. Assume $y^* \in M((-\infty, -\lambda - \mu - \epsilon])$. Then there exists a C^∞ function with support in $(\mu + \lambda, \infty)$, identical to 1 on $[\lambda + \mu + \epsilon, \infty)$, and such that $y \in [\int \alpha_a y f(a) \, da]$. Hence one finds $\varphi(x\alpha_a y) = 0$ which implies $E(\lambda + \mu + \epsilon)x\varphi = x\varphi$. Since $x\varphi E(\lambda) = x\varphi$ one sees that $x\varphi$ is an element of $\mathcal{A}_0^*(\mathbb{R}'^+)$. The other statement follows from passing to the adjoint.

(vii) From (vi) we know if $\varphi \in \mathcal{A}_0^*(\mathbb{R}'^+)$ and $x \in M([-\mu, \mu])$ the functional φx is again in $\mathcal{A}_0^*(\mathbb{R}'^+)$. Since $\mathcal{A}_0^*(\mathbb{R}'^+)$ is norm-dense in $\mathcal{A}^*(\mathbb{R}'^+)$, which itself is norm-closed, it follows that $\varphi \in \mathcal{A}^*(\mathbb{R}'^+)$ and $x \in M([-\mu, \mu])$ implies $\varphi x \in \mathcal{A}^*(\mathbb{R}'^+)$. It remains to show that for $x \in \mathcal{A}^{**}$ and $\varphi \in \mathcal{A}^*(\mathbb{R}'^+)$ there exists a sequence $x_n \in M([-\mu_n, \mu_n])$ such that φx_n converges in norm to φx. Notice first if $f \in \mathcal{L}^1(\mathbb{R})$ with support $\mathcal{F}^{-1}f \subset [-\mu, \mu]$ then $\varphi x(f) \in \mathcal{A}^*(\mathbb{R}'^+)$. Since these functions are norm-dense in $\mathcal{L}^1(\mathbb{R})$ it follows that $\varphi x(f) \in \mathcal{A}^*(\mathbb{R}'^+)$ for every $f \in \mathcal{L}^1(\mathbb{R})$. From the estimate

$$|\varphi(y[x(f) - x])| \leq \{|\varphi|(yy^*)|\varphi|([x(f)^* - x^*][x(f) - x])\}^{\frac{1}{2}} \qquad (*)$$

we obtain $(*) \leq \|y\| \, \|\varphi\|^{\frac{1}{2}} \, |\varphi|([x(f)^* - x^*][x(f) - x])^{\frac{1}{2}}$. It remains to show that the last expression converges to zero for a suitably chosen sequence f_n. Notice first that $\varphi E^+ = \varphi$ implies $E^+|\varphi| = |\varphi|E^+ = |\varphi|$ and hence $|\varphi|$ also belongs to $\mathcal{A}^*(\mathbb{R}'^+)$. Therefore, given ϵ then one can find δ_1 with $\||\varphi| \circ \alpha_a - |\varphi|\| < \frac{\epsilon}{4\|x\|^2}$ for $|a| < \delta_1$. Since $x^*\varphi \in \mathcal{A}_c^*$ there exists δ_2 with $\|(x^*|\varphi|) \circ \alpha_a - x^*|\varphi|\| < \frac{\epsilon}{4\|x\|}$ for $|a| < \delta_2$, and δ_3 with $\|(|\varphi|x) \circ \alpha_a - |\varphi|x\| < \frac{\epsilon}{4\|x\|}$ for $|a| < \delta_3$. Let now δ be such that $\delta < \delta_1, \delta < \frac{\delta_2}{2}$ and $\delta < \delta_3$ and assume $f \in \mathcal{L}^1(\mathbb{R})$ with supp $f \subset [-\delta, \delta]$ and $\int f(a) \, da = 1$. Under these assumptions one obtains the following estimate:

$$|\varphi|([x(f)^* - x^*][x(f) - x])$$
$$= |\varphi|(x(f)^* x(f)) - |\varphi|(x^* x(f)) - |\varphi|(x(f)^* x) + |\varphi|(x^* x)|$$
$$\leq |\int \{|\varphi|(x^* x) - |\varphi|(x^* \alpha_b x)\} f(b)\, db|$$
$$+ |\int \{|\varphi|(x^* x) - |\varphi|((\alpha_a x^* x)\} \overline{f}(a)\, da|$$
$$+ |\int\int \{|\varphi|(x^* \alpha_{b-a} x) - |\varphi|(x^* x)\} \overline{f}(a) f(b)\, da\, db|$$
$$+ |\int\int \{|\varphi| \circ \alpha_a(x^* \alpha_{b-a} x) - |\varphi|(x^* \alpha_{b-a} x)\} \overline{f}(a) f(b)\, da\, db|$$
$$\leq \frac{\epsilon}{4} \left(\int f(b)db + \int \overline{f}(a)da + \int\int \overline{f}(a) f(b)\, da\, db \right.$$
$$\left. + \int\int \overline{f}(a) f(b)\, da\, db \right) = \epsilon.$$

This shows that φx is the norm-limit of $\varphi x(f)$'s which belong to $\mathcal{A}^*(\mathbb{R}'^+)$. So $\varphi x \in \mathcal{A}^*(\mathbb{R}'^+)$. Since this space is $*$-invariant one has $y\varphi x \in \mathcal{A}^*(\mathbb{R}'^+)$ for $\varphi \in \mathcal{A}^*(\mathbb{R}'^+)$ and $x, y \in \mathcal{A}^{**}$.

(viii) E^+ is the common support projection of all $\varphi \in \mathcal{A}^*(\mathbb{R}'^+)$. This means $E^+ \in \mathcal{A}^{**}$. On the other hand since φx is also in $\mathcal{A}^*(\mathbb{R}'^+)$ for every $x \in \mathcal{A}^{**}$ it follows that E^+ commutes with every $x \in \mathcal{A}^{**}$, and hence $E^+ \in \mathcal{Z}(\mathcal{A}^{**})$ which implies that $\mathcal{A}^*(\mathbb{R}'^+)$ is a folium. □

Remark:

Since $E^+ \in \mathcal{Z}(\mathcal{A}^{**})$ it follows that $\mathcal{A}^*(\mathbb{R}'^+)$ is the set of normal states on $E^+ \mathcal{A}^{**}$ which is a von Neumann algebra. If $x \in \mathcal{A}^{**}$ then $E^+ \alpha_a x$ is weakly continuous; in other words we can write $x(f)$ if $x \in \mathcal{A}^{**} E^+$ instead of $[x(f)]$.

In $\mathcal{A}^{**} E^+$ the operator $\int_0^\infty e^{ia\lambda} dE(\lambda)$ is a one-parameter group of unitary operators and it is our aim to show that this implements the automorphism α_a.

For use in the proof of the next proposition we recall a consequence of Definition II.3.2. Let $f \in \mathcal{L}^1(\mathbb{R})$ with supp $\mathcal{F}^{-1} f \subset [\lambda, \mu]$. Then it follows that $[x(f)] \subset M([\lambda, \mu])$ provided one has $\lambda < \mu$.

II.4.5 Proposition:

With the same assumptions and notations as before one obtains

(i) *If $x \in E^+ M(D_1)$ and $y \in E^+ M(D_2)$ then it follows that*

$$xy \in M(\overline{D_1 + D_2}).$$

(ii) *For every $x \in \mathcal{A}^{**}$ one has $(\lambda > 0)$*

$$xE(\lambda) \in M([-\lambda, \infty)).$$

(iii) *For every $x \in \mathcal{A}^{**}$ one has $(\lambda, \mu > 0)$*

$$(E^+ - E(\lambda))xE(\mu) \in M([-\mu + \lambda, \infty)).$$

(iv) *For every $x \in \mathcal{A}^{**}$, and $0 < \lambda_1 < \lambda_2$ and $0 < \mu_1 < \mu_2$ one has*

$$\big(E(\lambda_2) - E(\lambda_1)\big)x\big(E(\mu_2) - E(\mu_1)\big) \in M([-\mu_2 + \lambda_1, -\mu_1 + \lambda_2]).$$

Proof: (i) $x \in M(D)$ is equivalent to the statement that supp $\mathcal{F}^{-1}\alpha_a x E^+ \subset D$. Hence one has thas $\mathcal{F}^{-1}\alpha_a(x)\alpha_b(y)E^+$ has its support in $D_1 \times D_2$. Denoting $(\mathcal{F}^{-1}\alpha_a(x)\alpha_b(y))(t_1, t_2) = F(t_1, t_2)$ one finds that $\alpha_a(x, y) = \int e^{ia(t_1 + t_2)} F(t_1, t_2) dt_1\, dt_2$ where the integral is meant in the distributional sense. But this shows $xy \in M(\overline{D_1 + D_2})$.

(ii) For $f \in \mathcal{L}^1(\mathbb{R})$ and supp $\mathcal{F}^{-1}f \in (-\infty, -\lambda]$ it follows that $E^+ x(f) \in M((-\infty, -\lambda])$ and hence $x(f)E(\lambda) = 0$ by the definition of $E(\lambda)$. This shows supp $\mathcal{F}^{-1}\alpha_a x E(\lambda) \subset [-\lambda, \infty)$ which is equivalent to the above statement.

(iii) Assume $y \in M((-\infty, -\lambda])$ and let f be as in (ii) such that $x(f) \in M((-\infty, -\mu + \lambda])$. Then from (i) one has $yx(f) \in M((-\infty, -\mu])$. Thus we obtain $yx(f)E(\mu) = 0$. From this we learn by (ii) supp $\mathcal{F}^{-1}\alpha_a yx E(\mu) \subset (-\infty, -\mu + \lambda]$ for every $y \in M((-\infty, -\lambda])$. Now $yx(f)E(\mu) = 0$ implies $zyx(f)E(\mu) = 0$ for every z in $\mathcal{A}^{**}$ and hence for the whole weakly closed left ideal generated by $M((-\infty, -\lambda])$. Since $E(\lambda)$ is its right annihilator it follows $1 - E(\lambda)$ belongs to this left ideal and hence $(E^+ - E(\lambda))x(f)E(\mu) = 0$ if $x(f) \in M((-\infty, -\mu + \lambda])$ or equivalently supp $\mathcal{F}^{-1}(E^+ - E(\lambda))\alpha_a x E(\mu) \in [-\mu + \lambda, \infty)$.

(iv) Let X be the expression (iv). This we can write in two ways:

$$\begin{aligned}
X &= (E^+ - E(\lambda_1))\{E(\lambda_2)x(E^+ - E(\mu_1))\}E(\mu_2) \\
&= E(\lambda_2)\{(E^+ - E(\lambda_1))xE(\mu_2)\}(E^+ - E(\mu_1)).
\end{aligned}$$

Hence we obtain $X \in M([\lambda_1 - \mu_2, \infty))$ and $X^* \in M([\mu_1 - \lambda_2, \infty))$. The desired inclusion is implied by Proposition II.3.3. □

After these preparations we are able to prove the main result of this section.

II.4.6 Theorem:

Let $\{\mathcal{A}, \mathbb{R}, \alpha\}$ be a C^*-dynamical system and assume the projection E^+ defined in II.4.1 is not zero, $E^+ \in \mathcal{Z}(\mathcal{A}^{**})$. Let $E(\lambda) \in \mathcal{A}^{**}E^+$ be the family of projections which are defined in II.4.1. Let

$$U(a) = \int_0^\infty e^{ia\lambda}\, dE(\lambda)$$

which is a unitary group in $\mathcal{A}^{**}E^+$. Then

(i) $U(a)$ *implements the automorphisms* α_a; *namely:*

$$\alpha_a x E^+ = U(a)xU^*(a)$$

for every $x \in \mathcal{A}^{**}$.

(ii) $U(a)$ *is minimal in the following sense. Let* $\{\pi, \mathcal{H}\}$ *be a normal representation of* $\mathcal{A}E^+$. *Assume on* $\mathcal{H}$ *there exists a continuous unitary group* $V(a)$ *such that*

(α)

$$\pi(\alpha_a x) = V(a)\pi(x)V^*(a).$$

(β) *The spectrum of* $V(a)$ *is contained in* $\mathbb{R}^+$.

If we write $U(a) = \exp\{iHa\}$ *and* $V(a) = \exp\{iH'a\}$ *then follows* $H' \geq \pi(H)$.

(Remark that the spectrum condition implies that H and H' are both non-negative operators).

Proof: Let λ_0 be fixed then $E(\lambda_0)\mathcal{A}^{**}E(\lambda_0)$ is an invariant subalgebra and our aim is to show that $E(\lambda_0)U(a)$ implements the automorphism on this algebra. Choose $0 = \mu_0 < \mu_1 < ... < \mu_N = \lambda_0$. Choose $\nu_i \in (\mu_{i-1}, \mu_i]$ and define

$$U_I(a) = \sum_{j=1}^{N} e^{ia\nu_j}\left(E(\mu_j) - E(\mu_{j-1})\right)$$

where I stands for the set $\{\mu_i, \nu_i\}$. Let $\epsilon(I) = \max(\mu_i - \mu_{i-1})$. Now we want to investigate the expression

$$U_I^*(a)\alpha_a x U_I(a) = \sum_{j,k} e^{-ia\nu_j + ia\nu_k}\left(E(\mu_j) - E(\mu_{j-1})\right)\alpha_a x\left(E(\mu_k) - E(\mu_{k-1})\right).$$

One obtains from Proposition II.4.6 (iv)

$$\text{supp } \mathcal{F}^{-1}e^{ia(\nu_k - \nu_j)}\left(E(\mu_j) - E(\mu_{j-1})\right)\alpha_a x\left(E(\mu_k) - E(\mu_{k-1})\right)$$
$$\subset [\mu_{j-1} - \mu_k, \mu_j - \mu_{k-1}] + \nu_k - \nu_j \subset [-\epsilon(I), \epsilon(I)].$$

This shows that $U_I^*(a)\alpha_a x U_I(a)$ has an extension as an entire analytic function of exponential type $2\epsilon(I)$ which is bounded for real a by $\|xE^+\|$. Thus we obtain $\|U_I^*(a)\alpha_a x U_I(a)\| \leq \|x\|e^{|\text{Im } a|2\epsilon(I)}$ and $\|U_I^*(a)\alpha_a x U_I(a) - x\| \leq 2|a|\epsilon(I)e^2$ for $|a\epsilon(I)| \leq 1$ and with $x \in E(\lambda_0)\mathcal{A}^{**}E(\lambda_0)$. Choosing a sequence I_n such that $\epsilon(I_n) \to 0$, $U_{I_n}(a)$ converges to $U(a)E(\lambda_0)$ in the norm-topology and this uniformly on every compact of $\mathbb{C}$. Consequently we find

$$U(a)xU(a)^* = \alpha_a x \quad \text{for} \quad x \in E(\lambda_0)\mathcal{A}^{**}E(\lambda_0).$$

Now notice that $E(\lambda)$ converges strongly to E^+ for $\lambda \to \infty$ and $E(\lambda)U(a)$ converges strongly to $U(a)$. Therefore, we obtain statement (i) of the theorem by taking the weak limit $\lambda \to \infty$ of

$$E(\lambda)U(a)xU(a)^*E(\lambda) = E(\lambda)\alpha_a x E(\lambda).$$

Note that there exists a projection $G_\pi \in \mathcal{Z}(\mathcal{A}^{**}E^+)$ such that $G_\pi \mathcal{A}^{**}$ and $\pi(\mathcal{A}^{**})$ are isomorphic. Let $V(a) = \int_0^\infty e^{ia\lambda} dE'(\lambda)$ be the spectral decomposition of $V(a)$ then $\pi(H) \leq H'$ is equivalent to the statement $E'(\lambda) \leq \pi(E(\lambda))$ for every λ. Since $E(\lambda)$ is the right annihilator of $M((-\infty, -\lambda])$ and since $\mathcal{A}^{**}$ is a von Neumann algebra it follows that $E(\lambda)G_\pi$ is the right annihilator of $G_\pi M((-\infty, -\lambda])$. Therefore, $E'(\lambda) \leq E(\lambda)$ holds if we have $\pi(x)E'(\lambda) = 0$ for every $x \in M((-\infty, -\lambda])$. Now we compute $V(a)\pi(x)E'(\lambda) = \pi(\alpha_a x)V(a)E'(\lambda) = \lim_{\epsilon \to 0} V(a)E'(\lambda - \epsilon)$. Since $V(a)$ has a positive spectrum one has supp $\mathcal{F}^{-1}V(a)E'(\lambda - \epsilon) \subset [0, \lambda - \epsilon]$. By assumption one knows supp $\mathcal{F}^{-1}\pi(\alpha_a x) \subset (-\infty, -\lambda]$ and hence supp $\mathcal{F}^{-1}\pi(\alpha_a x)V(a)E'(\lambda - \epsilon) \subset (-\infty, -\epsilon]$. But again $V(a)$ has a positive spectrum and this means supp $\mathcal{F}^{-1}V(a) \subset [0, \infty)$. This can only hold if $V(a)\pi(x)E'(\lambda - \epsilon) = 0$. Since $V(a)$ is unitary we have $\pi(x)E'(\lambda - \epsilon) = 0$ and hence $\pi(x)E'(\lambda) = 0$ which implies $E'(\lambda) \leq E(\lambda)$. □

II.5 Characterization of positive energy states

In this section we will first treat the converse problem, namely we again consider a C^*–dynamical system $\{\mathcal{A}, \mathbb{R}, \alpha\}$ and a representation $\{\mathcal{H}, \pi, V, \mathbb{R}'^+\}$ by which is meant a representation π of $\mathcal{A}$ on $\mathcal{H}$, and a continuous unitary representation $V(a)$ of the group $\mathbb{R}$ with spectrum in $\mathbb{R}'^+$ which implements the automorphism α_a. We want to investigate the relations between this representation and objects of the last section. Furthermore we give a characterization of those states which give rise to representations with the properties just described. We start with a general observation.

II.5.1 Lemma:

Let $\{\mathcal{A}, G, \alpha\}$ be a C^*–dynamical system. Let $G(\tau)$ be a topological group, and let $\{\mathcal{H}, \pi, U(g)\}$ be a representation of $\mathcal{A}$, where $U(g)$ a continuous representation of G implementing the automorphisms α, i.e.,

$$U(g)\pi(x)U^*(g) = \pi(\alpha_g x).$$

Then every normal functional of π belongs to $\mathcal{A}_c^*$ (see II.2.1 for the definition).

Proof: Since $\mathcal{A}_c^*$ is a vector space which is closed in the norm-topology and since every π-normal functional is the norm-limit of linear combinations of vector functionals it is sufficient to prove the statement for vector functionals. Now we compute

$$|(\varphi, \{\pi(\alpha_g x) - \pi(x)\}\psi)| = |(\varphi, \{U(g)\pi(x)U^*(g) - \pi(x)\}\psi)|$$
$$\leq |(\varphi, U(g)\pi(x)\{U^*(g) - 1\}\psi)| + |(\varphi, \{U(g) - 1\}\pi(x)\psi)|$$
$$\leq \|x\| [\|\varphi\| \|\{U^*(g) - 1\}\psi\| + \|\{U^*(g) - 1\}\varphi\| \|\psi\|]$$

from which we obtain:

$$\|(\varphi, .\psi) \circ \alpha_g - (\varphi, .\psi)\|$$
$$\leq [\|\varphi\| \|\{U^*(g) - 1\}\psi\| + \|\{U^*(g) - 1\}\varphi\| \|\psi\|].$$

Since $U(g)$ is continuous and hence strongly continuous it follows that $(\varphi, .\psi) \in \mathcal{A}_c^*$. □

Before going on we need a preparatory result:

II.5.2 Lemma:

Let $\{\mathcal{A}, \mathbb{R}^d, \alpha\}$ be a C^*–dynamical system. Let $D \subset \mathbb{R}^{\prime d}$ be a closed set. Let $\epsilon > 0$ and denote by D_ϵ an ϵ-neighbourhood of D. Denote by $R(D_\epsilon)$ the weak closure of the linear space generated by the sets $[x(f)]$ (see Definition II.3.1) with $x \in \mathcal{A}^{**}$ and supp $\mathcal{F}^{-1}f \subset D_\epsilon$. Then $R(D_\epsilon)$ contains $M(D)$ (see Definition II.3.2 for $M(D)$).

Proof: For every $x \in \mathcal{A}^{**}$ and every g with supp $\mathcal{F}^{-1}g \subset D_\epsilon$ one has $[x(g)] \subset R(D_\epsilon)$. If $x \in M(D)$ then one hase supp $\mathcal{F}^{-1}\alpha_a x \subset D$. Now for every function $f \in \mathcal{L}^1(\mathbb{R}^d)$ there exists a function $g \in \mathcal{L}^1(\mathbb{R}^d)$ with supp $\mathcal{F}^{-1}g \subset D_\epsilon$ and $\mathcal{F}^{-1}f = \mathcal{F}^{-1}g$ on D. Hence it follows that $[x(f)] \subset R(D_\epsilon)$ for every $x \in M(D)$ and every $f \in \mathcal{L}^1(\mathbb{R}^d)$. So it remains to approximate $[x]$ by $[x(f)]$. For $\varphi \in \mathcal{A}_c^*$ denote $\varphi_f = \int f(a)\varphi \circ \alpha_a \, da$ which is well defined since $\varphi \circ \alpha_a$ is continuous in the norm-topology. Using the identity $\varphi(x(f)) = \varphi_f(x)$ we see that the right-hand side is continuous in the ultraweak topology on $\mathcal{A}^{**}$. With this we find that if $\varphi \in \mathcal{A}_c^*$ then there exists a function $f \in \mathcal{L}^1(\mathbb{R}^d)$ with $|\varphi(x(f)) - \varphi(x)| < \epsilon$ (see the proof of Proposition II.4.4 (vii)) and hence $x \in R(D_\epsilon)$. □

We use this result for proving the important

II.5.3 Theorem:

Let $\{\mathcal{A}, \mathbb{R}, \alpha\}$ be a C^*–dynamical system. Then a functional $\varphi \in \mathcal{A}^*$ belongs to $\mathcal{A}_0^*(\mathbb{R}^{\prime +})$ if and only if φ and φ^* both have the following properties:

(α) The functions $a \to \varphi(x\alpha_a y)$, $x, y \in \mathcal{A}$ are continuous

(β) $\varphi(x\alpha_a y)$ is the boundary value of an analytic function $W_{x,y}(z)$ holomorphic in the upper half-plane $\{z; \text{Im } z > 0\}$

(γ) There exists a constant $m > 0$ such that $W_{x,y}(z)$ fulfils the estimate

$$|W_{x,y}(z)| \le \|\varphi\| \, \|x\| \, \|y\| e^{m\,\mathrm{Im}\,z}.$$

Proof. Assume first that $\varphi \in \mathcal{A}_0^*(\mathbb{R}'^+)$. Then there exists λ such that $\varphi = \varphi E(\lambda)$ and $\varphi^* = \varphi^* E(\lambda)$. From this we obtain the conditions as follows: $\varphi(x\alpha_a y) = \varphi(x\alpha_a(y)E(\lambda))$. Since $\varphi \in \mathcal{A}_0^*(\mathbb{R}'^+)$ we have $x\varphi \in \mathcal{A}^*(\mathbb{R}'^+)$ by Proposition II.4.4 (vii). On the other hand $yE(\lambda) \in M([-\lambda, \infty))$ by Proposition II.4.5 (ii) which means that $\varphi(x\alpha_a(y)E(\lambda))$ is the Fourier transform of a tempered distribution with support in $[-\lambda, \infty)$. Hence $\varphi(x\alpha_a y)$ has an analytic continuation $W_{x,y}(z)$ into the upper half-plane. $W_{x,y}(z)$ is of exponential type λ. On the real axis it is bounded by $\|\varphi\| \, \|x\| \, \|y\|$ and therefore we obtain the estimate $|W_{x,y}(z)| \le \|\varphi\| \, \|x\| \, \|y\| e^{(\lambda+\epsilon)\mathrm{Im}\,z}$. This shows that the conditions are fulfilled. Conversely assume φ is a functional fulfilling the three conditions. Putting $x = 1$ it follows from $\overline{\varphi(\alpha_a y)} = \varphi^*(\alpha_a y^*)$ that $\varphi(\alpha_a y)$ extends to an entire analytic function $W_y(z)$ which can be estimated as follows: $|W_y(z)| \le \|\varphi\| \, \|y\| e^{m|\mathrm{Im}\,z|}$. Hence we get by the Schwarz lemma

$$|\varphi(\alpha_a y) - \varphi(y)| \le |a| \, \|y\| 2\|\varphi\| e^m \quad \text{for} \quad |a| \le 1$$

which shows that $\varphi \circ \alpha_a$ is continuous in a. From the estimate (γ) it follows that $\varphi(x\alpha_a y)$ is the Fourier transform of a tempered distribution with support in $[-m, \infty)$. Hence for every $f \in \mathcal{L}^1(\mathbb{R})$ with $\mathrm{supp}\,\mathcal{F}^{-1}f \subset (-\infty, -m]$ we get $\int \varphi(x\alpha_a y)f(a)da = 0$. Since $x\varphi$ is a normal functional on $\mathcal{A}^{**}$ we find, by Lemma II.5.2, that $\varphi(xy) = 0$ for every $y \in M((-\infty, -m-\epsilon]), \epsilon > 0$ and consequently, by the definition of $E(\lambda)$, the equation $\varphi E(m+\epsilon) = \varphi$. Since φ^* has the same property we also get $E(m+\epsilon)\varphi = \varphi$, which shows that $\varphi \in \mathcal{A}_0^*(\mathbb{R}'^+)$. $\qquad\square$

Remark:

It is necessary to require that $\varphi(x\alpha_a y)$ and $\varphi(\alpha_a(x)y)$ be continuous since this cannot be concluded from the continuity of the function $\varphi(\alpha_a x)$.

Now we are prepared to prove the main result of this section.

II.5.4 Theorem:

Let $\{A, \mathbb{R}, \alpha\}$ be a C^*-dynamical system, and let $\{\mathcal{H}, \pi\}$ be a representation of A. Then the following statements are equivalent:

(1) In $B(\mathcal{H})$ there exists a continuous unitary representation $V(a)$ of the group $\mathbb{R}$ with

(α) spectrum $V(a) \subset \mathbb{R}'^+$.

(β) $V(a)$ implements the automorphism α_a:

$$\pi(\alpha_a x) = V(a)\pi(x)V^*(a).$$

(2) *There exists a projection $G_\pi \in \mathcal{Z}(\mathcal{A}^{**}E^+)$ such that $\mathcal{A}^{**}G_\pi$ and π'' are isomorphic.*

(3) *Every vector state of π belongs to $\mathcal{A}^*(\mathbb{R}'^+)$.*

Proof: We show (1) $\rightarrow$ (3) $\rightarrow$ (2) $\rightarrow$ (1). (1) $\rightarrow$ (3) Let $V(a) = \int_0^\infty e^{ia\lambda} dE'(\lambda)$ be the integral decomposition of $V(a)$ and take $\psi \in \mathcal{H}$ such that $E'(\lambda)\psi = \psi$ for some λ. Then we have:

$$(\psi, \pi(x)\pi(\alpha_a y)\psi) = (\psi, \pi(x)V(a)\pi(y)V^*(a)E'(\lambda)\psi) = \omega_\psi(x\alpha_a y).$$

Since spectrum $V(A) \subset \mathbb{R}'^+$ it follows that supp $\mathcal{F}^{-1}\omega_\psi(x\alpha_a y) \subset [-\lambda, \infty)$, which implies that $\omega_\psi(x\alpha_a y)$ has an analytic extension $W(z)$ into the upper half-plane as a function of exponential type λ. Since $W(z)$ is bounded for real z by $\|\psi\|^2\|x\|\,\|y\|$ one has the estimate $|W(z)| \le \|\psi\|^2\|x\|\,\|y\|e^{(\lambda+\epsilon)\text{Im}\,z}$. This shows that ω_ψ fulfils the conditions of Theorem II.5.3 and hence it belongs to $\mathcal{A}_0^*(\mathbb{R}'^+)$. Since the set of vectors with the property $E'(\lambda)\psi = \psi$ for some λ is dense in $\mathcal{H}$ we get that every ω_ψ is the norm-limit of these special ω'_ψs. Hence every ω_ψ belongs to $\mathcal{A}^*(\mathbb{R}'^+)$ since this is a norm-closed linear space (by Prop. II.4.4).

(3) $\rightarrow$ (2) Since all the vector states of π belong to $\mathcal{A}^*(\mathbb{R}'^+)$ and since this is a norm-closed linear space all normal states of π belong to $\mathcal{A}^*(\mathbb{R}'^+)$. This implies that there exists a projection $G_\pi \in \mathcal{Z}(\mathcal{A}^{**}E^+)$ such that a positive functional is π-normal if $\omega(G_\pi) = \|\omega\|$ holds. But this implies that π'' and $\mathcal{A}^{**}G_\pi$ are equivalent von Neumann algebras.

(2) $\rightarrow$ (1). Let $U(a) \in \mathcal{A}^{**}E^+$ be the unitary group implementing α_a which has been constructed in the last section. Then for $x \in \mathcal{A}^{**}E^+$ one has $U(a)xU^*(a) = \alpha_a x$ and since π is quasi-equivalent to a subalgebra of $\mathcal{A}^{**}E^+$ one obtains also the equation:

$$\pi\big(U(a)\big)\pi(x)\pi\big(U(a)\big)^* = \pi(\alpha_a x).$$

Since $\pi\big(U(a)\big)$ is again unitary (1) follows. $\qquad\qquad\square$

II.6 The spectrum in a cone

The following situation will be investigated in this section. $\{\mathcal{A}, \mathbb{R}^d, \alpha\}$ is a C^*-dynamical system and $C \subset \mathbb{R}^d$ is a cone with the properties

(i) $\overline{C}$ is a closed convex cone.

(ii) C is an open cone.

(iii) $\overline{C}$ is a proper cone i.e. $\overline{C} \cap \{-\overline{C}\} = \{0\}$.

The dual cone will be denoted by C' which is a proper closed cone with interior points. We want to characterize covariant representations $\{\mathcal{H}, \pi, U(a)\}$ such that $U(a)$ is a continuous unitary representation of the group $\mathbb{R}^d$ with spectrum $U(a)$ contained in C', and the property that it implements the automorphisms α_a. This means in particular that for every direction $t \in \overline{C}$ the subgroup $\alpha_{\rho t}$, $\rho \in \mathbb{R}$ fulfils the conditions of section II.4. This leads to the following

II.6.1 Definition:

Let $\{A, \mathbb{R}^d, \alpha\}$ be a C^*-dynamical system and let $\overline{C} \subset \mathbb{R}^d$ be a proper convex closed cone with interior points:

(i) For $t \in \overline{C}$, $t \neq 0$ denote by $E(t, \lambda)$ the spectral projection of the group representation $U(\rho t)$, $\rho \in \mathbb{R}$, described in section II.4. These projections are invariant under α_a by Proposition II.4.2(v).

(ii) By $E^+(t)$ we denote s-lim$_{\lambda \to \infty}$ $E(t, \lambda)$ which belongs to $\mathcal{Z}(A^{**})$ by II.4.4 (viii).

(iii) Define $E(C') = \prod\{E(t)^+; \, t \in \overline{C}, \, t \neq 0\}$ where the product is the limit of the decreasing net of finite products. $E(C') \in \mathcal{Z}(A^{**})$.

(iv) For $p \in C'$ define

$$E(<0, p>) = \prod\{E(t, \lambda_t); \, t \in \overline{C}, \, t \neq 0, \, \lambda_t = (p, t)\},$$

in which $< 0, p >$ stands for the order interval $C' \cap \{p - C'\}$. These projections are again invariant under α_a.

We start the investigation with the following preparation:

II.6.2 Lemma:

With the notation of Definition II.6.1 we obtain:

(i) Let p_n be increasing relative to the order of C' such that $\bigcup_n < 0, p_n >$ covers all of C'. Then one finds s-lim$_{n \to \infty}$ $E(< 0, p_n >) = E(C')$.

(ii) For every $x \in A^{**}$ the function

$$a \to \alpha_a(x)E(<0, p>)$$

is weakly continuous and

$$\mathrm{supp}\mathcal{F}^{-1}\alpha_a(x)E(<0, p>) \subset -p + C'.$$

Proof: (i) Since all the $U(\rho t)E(C')$ are continuous and commute for different t, it follows that every normal state of $A^{**}E(C')$ belongs to A_c^*. Let ω be a normal state of $A^{**}E(C')$ and let $f \in \mathcal{L}^1(\mathbb{R}^d)$ with $f \geq 0$, $\int f(a)\mathrm{d}a = 1$ with $\mathrm{supp}\mathcal{F}^{-1}f = K$ compact. Define $\omega_f = \int \omega \circ \alpha_a f(a)\mathrm{d}a$. Let $t \in \overline{C}$. Then there exists λ_t^1 such that $K \subset \{p; (p, t) \geq -\lambda_t^1\}$. By Definition II.3.2 such

ω_f annihilate every $x \in M(\{p; (p,t) \le -\lambda_t^1\})$ and therefore we have by Definition II.4.1 $\omega_f(E(t, \lambda_t^1)) = \omega_f(E(C')) = 1$. Let now $p_n \in C'$ be such that $(p_n, t) \ge \lambda_t^1$. Then $\omega_f(\prod_{i=1}^n E(t_i, (p_n, t))E(C')) = \omega_f(E(C'))$, and hence by the definition of $E(< 0, p_n >)$ we obtain $\omega_f(E(< 0, p_n >)) = \omega_f(E(C'))$. Denote by $F = \text{s-} \lim_{n \to \infty} E(< 0, p_n >)$. Then we have $\omega_f(F) = \omega_f(E(C'))$ for every positive $\mathcal{L}^1(\mathbb{R}^d)$ function of norm 1. But since α_a acts strongly continuously on the pre-dual of $\mathcal{A}^{**}E(C')$ it follows that these states are dense in this space and hence the above equation holds for every $\mathcal{A}^{**}E(C')$ normal state. From this we obtain $F \ge E(C')$. The opposite inclusion is trivial.

(ii) We know from Proposition II.4.5 that supp $\mathcal{F}^{-1}\alpha_{\lambda t}(x)E(t, \mu) \subset \{p; (p,t) \ge -\mu\}$. Hence we obtain by the definition of $E(< 0, p >)$

$$\text{supp} \mathcal{F}^{-1}\alpha_a(x)E(< 0, p_0 >) \subset \cap_{t \in \overline{C}}\{p; (p,t) \ge -(p_0, t)\} = -p_0 + C'. \quad \square$$

With help of this Lemma one is able to construct a group representation fulfilling the spectrum condition. To this end we introduce the following

II.6.3 Definition:

With the same assumptions and notations as before:

(i) *$B = \{b^1, ..., b^d\}$ denotes a basis of $\mathbb{R}^d$ such that $b^i \in \overline{C}$, and $\{b^i\}$ are linearly independent.*

(ii) *For every $b^i \in B$ let $U(\lambda b^i) \in \mathcal{A}^{**}E(C')$ be the minimal representation of the one-parameter group fulfilling the spectrum condition and implementing the automorphisms $\alpha_{\lambda b^i}$ described in section II.4. For $a \in \mathbb{R}^d$ and $a = \sum \lambda_i b^i$ define*

$$U_B(a) = \prod_{i=1}^d U(\lambda_i b^i).$$

(iii) *By C'_B we denote the following set:*

$$C'_B = \{p; (p, b^i) \ge 0, \ i = 1, ..., d\}.$$

By construction of $U_B(a)$ it follows that spectrum $U_B(a) \subset C'_B$.

The representation $U_B(a)$ is certainly not the only one which implements the automorphisms α_a on $\mathcal{A}^{**}E(C')$. If $Z(a)$ is a continuous representation of translation-group $Z(a) \in \mathcal{Z}(\mathcal{A}^{**}E(C'))$ then $U_B(a)Z(a)$ again belongs to $\mathcal{A}^{**}E(C')$ and implements α_a. Therefore it is natural to ask whether or not one can find $Z(a)$ is such a way that the spectrum of $U_B(a)Z(a)$ is contained in C'. That one can find such a representation $Z(a)$ will be the next result.

II.6.4 Theorem:

Let $\{\mathcal{A}, \mathbb{R}^d, \alpha\}$ be a C^*-dynamical system and assume that $E(C')$ is not zero (for the definition see II.6.1 (iii)). Then there exists a continuous unitary representation $U(a)$ of the translation group with

(i) $U(a) \in \mathcal{A}^{**} E(C')$.

(ii) *Spectrum* $U(a) \subset C'$.

(iii) $U(a)$ *implements the automorphism* α_a *on* $\mathcal{A}^{**} E(C')$.

Proof: Let Γ be a compact set in C'_B and denote by $\Delta = <0, p> \subset C'$ which is also compact. Denote by $F(\Gamma)$ the spectral projections of $U_B(a)$ and let $E(\Delta)$ be the projections introduced in II.6.1 (iv). All these projections belong to $\mathcal{Z}(\mathcal{A}_G^{**})$ and hence $F(\Gamma)E(\Delta)$ is again an invariant projection. Taking a sequence Γ_n such that $\cup_n \Gamma_n = C'_B$ and a sequence $\Delta_n = <0, p_n>$ such that $\cup_n \Delta_n = C'$, one finds, by construction of $F(\Gamma)$ and by Lemma II.6.2 (i), that s-$\lim_{n\to\infty} F(\Gamma_n)E(\Delta_n) = E(C')$. Let $G(\Gamma, \Delta)$ be the central carrier of $F(\Gamma)E(\Delta)$, Then $G(\Gamma, \Delta)$ is nothing else but the common range projection of all elements of the form $xF(\Gamma)E(\Delta)$, $x \in \mathcal{A}^{**}$. Hence investigating the expression $G(\Gamma, \Delta)U_B(a)$ is the same as investigating the family of expressions

$$U_B(a)xE(\Delta)F(\Gamma) = \alpha_a(x)U_B(a)E(\Delta)F(\Gamma)$$
$$= \alpha_a(x)E(\Delta)U_B(a)F(\Gamma).$$

From the definition of $F(\Gamma)$ it follows that supp $\mathcal{F}^{-1}U_B(a)F(\Gamma) \subset \Gamma$ and from Lemma II.6.2 (ii) we know supp $\mathcal{F}^{-1}\alpha_a(x)E(\Delta) \subset -\Delta + C'$. Putting the two together one has supp $\mathcal{F}^{-1}U_B(a)\alpha_a(x)E(\Delta)F(\Gamma) \subset \Gamma - \Delta + C'$. Since Γ and Δ are both compact we can choose $q = q(\Gamma, \Delta)$ such that $\Gamma - \Delta + q(\Gamma, \Delta) \subset C'$. Hence spec $G(\Gamma, \Delta)U_B(a)e^{i(a, q(\Gamma, \Delta))} \subset C'$. Now take a sequence Γ_n, Δ_n covering C'_B and C'. Then $G(\Gamma_n, \Delta_n)$ converges to $E(C')$. Define

$$W(a) = \sum_1^\infty \{G(\Gamma_{n+1}, \Delta_{n+1}) - G(\Gamma_n, \Delta_n)\}e^{i(a, q(\Gamma_{n+1}, \Delta_{n+1}))}.$$

Then $W(a)$ is a continuous unitary representation of the translations belonging to $\mathcal{Z}(\mathcal{A}^{**}E(C'))$ and therefore $U(a) = U_B(a)W(a)$ is a continuous unitary representation of the translations belonging to $\mathcal{A}^{**}E(C')$ which implements α_a. It remains to show that $U(a)$ fulfils the spectrum condition. We have

$$U(a) = U_B(a)W(a) =$$
$$\sum_1^\infty \{G(\Gamma_{n+1}, \Delta_{n+1}) - G(\Gamma_n, \Delta_n)\}U_B(a)e^{i(a, q(\Gamma_{n+1}, \Delta_{n+1}))}.$$

Since by the above calculation and the definition of $q(\Gamma, \Delta)$ each term of the summand has its spectrum in C', it follows that spectrum $U(a) \subset C'$. □

Finally we want to characterize the normal states of the algebra $\mathcal{A}^{**}E(C')$. We start with some notations.

II.6.5 Definition:

With the notations of this section we set:

(i) $\mathcal{A}^*(C') = \{\varphi \in \mathcal{A}^*; E(C')\varphi = \varphi E(C') = \varphi\}$.
(ii) $\mathcal{A}_0^*(C') = \{\varphi \in \mathcal{A}^*;$ there exists $p \in C'$ with $E(<0, p>)\varphi = \varphi E(<0, p>) = \varphi\}$.

With these notations one obtains the following result:

II.6.6 Theorem:

Let $\{\mathcal{A}, \mathbb{R}^d, \alpha\}$ be a C^*–dynamical system and $\overline{C} \subset \mathbb{R}^d$ a closed, convex, proper cone with interior $\overset{\circ}{C} \neq \emptyset$. One obtains, in the notation of the last definition:

(i) *$\mathcal{A}_0^*(C')$ is norm–dense in $\mathcal{A}^*(C')$.*
(ii) *An element $\varphi \in \mathcal{A}^*$ belongs to $\mathcal{A}_0^*(C')$ if and only if it fulfils the following properties:*

    (α) *$a \to \varphi(x\alpha_a x)$ is a continuous function on $\mathbb{R}^d$, $x, y \in \mathcal{A}$.*
    (β) *$\varphi(x\alpha_a y)$ is the boundary value of an analytic function $W(z)$ holomorphic in the tube*

$$T(C) = \{z \in \mathbb{C}^d; \text{ Im } z \in \overset{\circ}{C}\}.$$

    (γ) *There exists a constant m such that*

$$|W(z)| \leq \|\varphi\| \|x\| \|y\| e^{m\|\text{Im } z\|}$$

    holds for $z \in T(C)$.
    (δ) *φ^* fulfils the same conditions as φ.*

(iii) *Let $\{\mathcal{H}, \pi\}$ be a representation of $\mathcal{A}$. Then one can find a continuous unitary representation $V(a)$ acting on $\mathcal{H}$, which implements α_a with spectrum $V(a) \subset C'$ if and only if every vector state ω_ψ belongs to $\mathcal{A}^*(C')$.*

Proof: (i) From Lemma II.6.2 (i) we know that $E(<0, p_n>)$ converges to $E(C')$. From this the statement follows in the same manner as in the proof of II.4.4 (ii).

(ii) Let $\varphi \in \mathcal{A}_0^*(C')$ such that $\varphi E(<0, p>) = \varphi$. Then from Lemma II.6.2(ii)

$$\text{supp } \mathcal{F}^{-1}\varphi(x\alpha_a y) \subset -p + C'.$$

This implies that $\varphi(x\alpha_a y)$ is the boundary value of an analytic function $W(z)$ holomorphic in $T(\overset{\circ}{C})$ fulfilling the estimate

$$|W(z)| \leq \|\varphi\| \, \|x\| \, \|y\| e^{\|p\| \|\operatorname{Im} z\|}.$$

Hence conditions $(\alpha)-(\delta)$ are fulfilled. Conversely, assume that φ fulfils these conditions. Then there exists a point $p_1 \in C'$ such that $< 0, p_1 > \supset \{p \in C'; \|p\| \leq m\}$. So we see that $W(z)e^{i(p_1, z)}$ is bounded and therefore by Lemma II.1.2 supp $\mathcal{F}^{-1}\varphi(x\alpha_a y) \subset -p_1 + C'$. From this it follows that φ annihilates the left ideal generated by $M(\{p; (p, t) \leq -(p_1, t) - \epsilon\})$ for $t \in \overline{C}$, which implies $\varphi E(t, (p_1, t) + \epsilon) = \varphi$. Since this holds for every $t \in \overline{C}$ we obtain $\varphi E(< 0, (1 + \epsilon)p_1 >) = \varphi$. As the same is true for φ^* we find $\varphi \in A_0^*(C')$.

(iii) Let $\{\mathcal{H}, \pi\}$ be a representation of A such that the vector states ω_ψ belong to $A^*(C')$. Since $A^*(C')$ is a norm-closed vector space it follows that every π-normal state belongs to $A^*(C')$. Hence there exists a projection G_π in $\mathcal{Z}(A^{**}E(C'))$ such that π is a faithful normal representation of $A^{**}G_\pi$. Hence $\pi(U(a))$ is the desired representation of the translations on $\mathcal{H}$. Here $U(a)$ is a representation of the translations described in Theorem II.6.4. On the other hand let $\{\mathcal{H}, \pi, V(a)\}$ be a covariant representation fulfilling the described properties. Let

$$V(a) = \int_{C'} e^{i(p, a)} \mathrm{d}F(p)$$

be the integral representation. Then for every $\psi \in \mathcal{H}$ with compact support, i.e., $F(\Delta)\psi = \psi$ where Δ is a compact subset of C', the function

$$\omega_\psi(x\alpha_a y) = (\psi, \pi(x)V(a)\pi(y)V^*(a)F(\Delta)\psi)$$

is such that supp $\mathcal{F}^{-1}\omega_\psi(x\alpha_a y) \subset -\Delta + C'$. This implies that ω_ψ fulfils the conditions $(\alpha)-(\delta)$ of statement (ii). For condition (δ) notice that ω_ψ is selfadjoint and hence ω_ψ belongs to $A^*(C')$. Since the vectors $\psi \in \mathcal{H}$ with $F(\Delta)\psi = \psi$ for some compact $\Delta \subset C'$ are dense in $\mathcal{H}$, and since $A^*(C')$ is closed in norm, every ω_ψ belongs to $A^*(C')$. □

We end this section with some remarks on the topology of the space of states in $A^*(C')$.

II.6.7 Addendum:

(1) *Denote by $S(C')$ the set of states belonging to $A^*(C')$. This space is sequentially complete.*

(2) *Denote by $S(m)$ the set of states $\omega \in S(C')$ having exponential growth of order m, i.e.*

$$S(m) = \{\omega \in S(C'); |\omega(\alpha_z x)| \leq \|x\| e^{m\|\operatorname{Im} z\|}\}.$$

The continuity condition (ii, α) *of Theorem* II.6.6 *prevents this set of being weakly compact. But one has*

$\mathcal{S}(m)$ *is closed in the topology of uniform convergence of all functions*

$$a \longrightarrow \omega(x\alpha_a y), \quad x, y \in \mathcal{A}$$

on compact sets of $\mathbb{R}^d$.

Proof: (1) The space $\mathcal{A}^*(C')$ is the pre-dual of the von Neumann algebra $E(C')\mathcal{A}^{**}$. But it is well known that the set of normal states of a von Neumann algebra is sequentially complete.

(2) The necessity of the condition is clear, since otherwise the continuity of the limit is not guaranteed. Since all functions $\omega_\beta(x\alpha_a y)$ have analytic continuation $W_\beta(z)$ into the tube $T(C)$ which are bounded by $\|x\| \, \|y\| e^{m\|\mathrm{Im}\, z\|}$, it follows from Theorem II.1.7 that these are Fourier transforms of tempered distributions with support in $K_m + C'$, where K_m is the ball of radius m. Since the expressions $\omega_\beta(x\alpha_a y)$ converge as functions wich are uniformly bounded, they converge as tempered distributions. Hence the limit is again the Fourier transform of a tempered distribution with support in $K_m + C'$. Hence also the limiting function has an analytic continuation $W(z)$ into $T(C)$ fulfilling the same estimate. Hence by Theorem II.6.6 the limiting functional belongs to $\mathcal{A}_0^*(C')$, and the estimate shows that it belongs to $\mathcal{S}(m)$. $\quad\square$

II.7 Notes and remarks

(1) That there exist close relations between support properties of functions and analyticity properties of their Fourier transforms has been know for some time. The classic text book on this subject is that of Paley and Wiener [[PW]]. For applications in physics the extension of this theory to tempered distributions is very useful. The famous Paley-Wiener-Schwarz theorem which describes those distributions with compact support by means of their Fourier transformations already appears in the textbook of L. Schwartz on distribution theory [[Schw]].

Lemma II.1.2 is due to L. Schwartz [Schw52]. His proof differs from the proof given here. It is not necessary to use the sharp cutting property in our proof. One obtains the same result by using a decomposition with the help of smeared–out step functions which lead to the same arguments. The existence of sharp cutting used in our proof is a simple application of the Hahn-Banach theorem. This case is easy to handle, since here the cutting is by a hyperplane. For the cutting along more complicated hypersurfaces see e.g. the exposé of Malgrange in [Mal].

For the detailed description of distributions with values in some other vector space used in II.1.4 we refer to the paper of L. Schwartz [Schw58].

The proof of Theorem II.1.5 uses unpublished ideas of L. Gårding which are based on methods developed by Köthe [Kö] and Tillmann [Ti]. Our Theorem II.1.7 is also closely related to the so-called

Bros-Epstein-Glaser Lemma:

Let $C \subset \mathbb{R}^d$ be a closed, convex, proper cone with interior points. Let $T \in \mathcal{S}'(\mathbb{R}^d)$ with supp $T \subset C$. Then there exists a continuous, polynomially bounded function G with supp $G \subset C$ and a polynomial $P(x)$, $x \in \mathbb{R}^d$ with

$$T = P(D)G.$$

For the proof of this result in special cases see Bros, Epstein, Glaser [BEG], for the general case see e.g. Reed and Simon Vol. II [[RS]].

(2) If $\{\mathcal{A}, G, \alpha\}$ is a C^*–dynamical system then it is a natural task to characterize covariant representations. There is even the book of Pederson [[Ped]] which is mainly dedicated to this question. In most of the investigations it is assumed that the group acts strongly continuously on the algebra, which means the function $g \to \alpha_g(x)$ is continuous in the norm-topology for every element $x \in \mathcal{A}$. Moreover, it is generally assumed that G is a locally compact group. In this case one may exploit integration theory on the group. If both assumptions are fulfilled then one can construct the crossed product between the algebra $\mathcal{A}$ and the group G. This is a C^*–algebra containing all information concerning covariant representations. The crossed product with a discrete group has been introduced for the first time by Turumaru [Tu]. The general case is due to Doplicher, Kastler, and Robinson [DKR]. However, it has been observed by the author that the existence of covariant representation is governed by the continuity of the group action in $\mathcal{A}^*$ and not in $\mathcal{A}$. In the case of locally compact groups it has been shown in [Bch69,73a] and for the general situation in [Bch83]. This result reads:

Theorem:

Let $\{\mathcal{A}, G, \alpha\}$ be a C^*–dynamical system, with G a topological group, and let π be a representation of $\mathcal{A}$. There exists a representation $\{\hat{\pi}, U(g)\}$ with $\hat{\pi}$ quasi-equivalent to π, and $U(g)$ is a continuous unitary representation of the group G implementing α_g, if and only if the folium $F(\pi)$ (of π normal states) is invariant under α_g^* and α_g^* acts strongly continuously on the folium $F(\pi)$.

The main result of section II.2 is Theorem II.2.3 which has been proved in [Bch83]. The proof presented here is due to P. Junglas. It is much shorter then the original one in [Bch83].

It has been shown in [Bch93] that $\mathcal{A}_c^*$ is the predual of a von Neumann algebra. This result uses the Tomita-Takesaki theory [[Ta70]]. The first step in this direction can be found in [Bch83]. The result is

Theorem:

Let $\{A, G, \alpha\}$ be a C^*-dynamical system, with G a topological group. Then A_c^* is the predual of a von Neumann algebra $\mathcal{M}_c \subset A^{**}$. This von Neumann algebra coincides with $\mathcal{M}_m^{(\mathrm{red})}$ where

$$\mathcal{M}_c = \{A \in A^{**}; \, A\omega \in A_c^*, \, \omega A \in A_c^*, \, \forall \omega \in A_c^*\}.$$

The suffix "(red)" means division by the anihilator of A_c^*.

(3) The concept of momentum transfer is common knowledge to all physicists. It has been used in several branches of physics, in particular scattering theory. In mathematical literature the spaces of momentum transfer are known under the name of spectral subspaces (of the automorphism group). They have been introduced by Godement [Go] and systematically studied by Arveson [Arv]. In some investigations the spaces of momentum transfer are replaced by the left ideal generated by these subspaces [Bch70a]. The main contribution of Arveson in this field is a clarification of the different roles played by the different possibilities of defining spaces of momentum transfer. The definition used here is a simple generalization of Arveson's definition to the case where the group does not act strongly continuouity on the algebra.

(4) The main breakthrough in the theory of automorphisms of a C^*-algebra was the result of Kadison [Kad] and Sakai [Sa] stating that every derivation of a C^*-algebra can be realized with help of the commutant with some operator belonging to the enveloping von Neumann algebra. Shortly afterwards the authors [Bch66] could generalize this result to one-parameter automorphism groups with semi-bounded spectrum, provided the group action is strongly continuous. From here to the result presented in section II.4 took almost two decades [Bch84]. The main difficulty consisted of finding the right definition for $A^*(\mathbb{R}'^+)$ and showing that it is a folium. The construction of the unique minimal $U(a)$, $a \in \mathbb{R}$, is due to Olesen and Pedersen [OP] and Olesen [Ol] in cases where the action of the group is continuous in norm, or strongly continuous on the algebra. The extension to the case treated here is straightforward.

(5) In case the action of the group is strongly continuous, the states giving rise to positive energy representations such that their presenting vector has only finite energy support, can easily be characterized by left ideals [Bch70b] and also [Bch73b]. This also shows that in this continuous case the states with fixed finite energy form a weak $(\sigma(A^*, A))$ compact set. This result is not longer true when the continuity assumption on α_a is dropped. A characterization of such states in the general situation was first developed in [Bch84].

(6) The step from the one-dimensional case to the n-dimensional situation was easy in the case where the cone in question was a simplical cone (see e.g. [Ol]). The trick passing from here to the case where the cone is a general has been developed in [Bch84]. The observation that the spectrum condition leads to weakly inner automorphisms did not come completely unexpectedly since Araki [Ar64] had observed, already in 1964, that the center of every representation obeying the spectrum condition is left pointwise fixed by the translations.

The result about the sequential completenes of the normal functionals of a von Neumann algebra can be found in [[Ta79]].

Chapter III

The Opposite Edge of the Wedge Problem

In chapter II we have seen that the theory of analytic functions plays an important role when one wants to characterize support properties in momentum space by properties in configuration space. For the further development of the theory of local observables one has to look at the interplay between the spectrum condition in momentum space and the locality condition in configuration space. The main tool for treating such problems has turned out, so far, to be the theory of analytic functions. This chapter is devoted to a presentation of these techniques.

In the first section we will give a short introduction to the theory of several complex variables. The statements will be given without proofs. We also collect only those results which are useful for our purpose. The content of these sections is centered around the so-called edge of the wedge problem. In the second section we give a proof of the edge of the wedge theorem. It deals with the situation where two domains have common boundary-points but otherwise an empty intersection.

In $\mathbb{C}^n$ not every domain G is a natural domain. In such a situation every function holomorphic in G can be analytically continued into a larger domain. Sections 3 to 7 deal with this situation. In section 3 we prove the so-called double cone theorem. In section 4 we present the Jost-Lehmann-Dyson theorem. This is a far-reaching result, but it is restricted to problems which appear typically in physics. In the final three sections we use the Jost-Lehmann-Dyson result as a starting point and try to obtain some results which go beyond the Jost-Lehmann-Dyson representation. For $z \in \mathbb{C}^n$ we write $z = x + iy$ with $x, y \in \mathbb{R}^n$.

III.1 Holomorphic functions of several complex variables

Due to locality in configuration space and the spectrum condition in momentum space, many functions appearing in quantum field theory are boundary values of analytic functions. As a result, the theory of analytic functions is a tool which has been used with great success and which should be used also in future investigations.

A domain $G \subset \mathbb{C}^n$ is called a natural domain whenever to every point $z_0 \in \partial G$ there exists a function $f(z)$ holomorphic in G which cannot be analytically continued into a complex neighbourhood of this point z_0. Since in $\mathbb{C}^1$ to every z_0 there exists the function $\frac{1}{z - z_0}$, which is singular at z_0, it follows that every domain in $\mathbb{C}^1$ is a natural one. This is no longer true in the case of more than one variable. This can best be demonstrated by an example:

III.1.1 Example:

Let B be the set $B = \{z \in \mathbb{C}^n; \|z\| := \left(\sum |z_i|^2\right)^{1/2} \leq 1\}$. Let H_1^+ be the set $H_1^+ = \{z \in \mathbb{C}^n; \operatorname{Im} z_1 > 0\}$ and define $G = H_1^+ \setminus B$. Then for $n > 1$ one finds that every function $f(z)$ holomorphic in G has an extension into H_1^+.

Proof: Define

$$F(z_1, \ldots z_n) = \frac{1}{2\pi i} \int\limits_{|\varphi|=2} \frac{f(z_1, \ldots, z_{n-1}, \varphi)}{\varphi - z_n} \, d\varphi.$$

Since we have chosen $|\varphi| > 1$ one sees that the integral is always defined for $z \in H_1^+$. Moreover, $F(z)$ is an analytic function in all variables. If now $\operatorname{Im} z_1 > 1$ then $\{z; \operatorname{Im} z_1 > 1\} \times \mathbb{C}^{n-1} \subset G$ and hence one obtains $F(z_1, \ldots, z_n) = f(z_1, \ldots, z_n)$ for such points. So $F(z_1, \ldots, z_n)$ is an analytic continuation of f. $\qquad\square$

This example shows that in special situations every function, analytic in G, can be extended into a larger domain. Doing all extensions possible one might come to a situation where one obtains a domain which is a covering space over some parts of $\mathbb{C}^n$. In such a case one is speaking of a Riemann surface. These considerations lead to the following

III.1.2 Definition:

(a) *Let G be a domain in $\mathbb{C}^n$. Then we denote by $H(G)$ the smallest domain into which every function, holomorphic in G, can be analytically extended. $H(G)$ might be a Riemann surface over $\mathbb{C}^n$. $H(G)$ is called the envelope of holomorphy of G.*

(b) *The envelope of holomorphy of a domain is called "schlicht" if $H(G)$ is again a domain in $\mathbb{C}^n$.*

If $G = H(G)$ then to every boundary-point there must exist a function which cannot be extended into a neighbourhood of this point implying that $H(G)$ is a natural domain in the above sense.

In $\mathbb{C}^1$ every domain G is a domain of holomorphy which implies that $G_1 \times G_2 \times ... \times G_n \subset \mathbb{C}^n$ are also domains of holomorphy. In the above example $H_1^+ \setminus B$ we have seen that every function can be extended into H_1^+. But H_1^+ is a domain of holomorphy $H_1^+ = H^+ \times \mathbb{C}^{n-1}$, so one obtains $H(H_1^+ \setminus B) = H_1^+$.

The above example also shows that one finds the envelope of holomorphy of G by cutting all "noses" which are sticking into the domain G. The only difficulty consists in recognizing these noses. They can have complicated shapes because the typical structure of such a "nose" might only be realized after some bi-holomorphic transformation. Some of the properties of domains of holomorphy will be mentioned here without proof.

III.1.3 Lemma:

(a) *Let G_1, G_2 be such that $G_1 \subset \mathbb{C}^n$, and $H(G_1) = G_1$ and $G_2 \subset \mathbb{C}^m$; $H(G_2) = G_2$. Then*

$$H(G_1 \times G_2) = G_1 \times G_2 \subset \mathbb{C}^{n+m}.$$

(b) *Let I be an index set and assume for $\alpha \in I$ one has $G_\alpha \subset \mathbb{C}^n$ with $H(G_\alpha) = G_\alpha$. Define $G = \{\bigcap_{\alpha \in I} G_\alpha\}^\circ$. Then it follows that*

$$H(G) = G.$$

(c) *Let $a_i, b \in \mathbb{C}$ with $\sum |a_i| \neq 0$; then*

$$\{z \in \mathbb{C}^n; \operatorname{Im}\left(\sum a_i z_i + b\right) > 0\}$$

is a domain of holomorphy.

(d) *Using (b) and (c) one finds that every convex open set in $\mathbb{C}^n$ is a domain of holomorphy.*

If f is an entire analytic function then it has a power series expansion, which implies that the polynomials form a dense subset of the set of all entire analytic functions. A corresponding result is true for the functions holomorphic in a natural domain. In order to describe this we set

$$d(z) = \text{distance of} \quad z \quad \text{from} \quad \mathbb{C}^n \setminus G$$

and

$$\Delta(z) = \min\{1, d(z), \|z\|^{-1}\}.$$

Then the following result holds (without proof):

III.1.4 Theorem:

Define $\Delta(z)$ as above, and assume $G = H(G)$. Then the set of functions $f(z)$ holomorphic in G fulfilling the estimate

$$|f(z)| \leq c\Delta^{-n}(z)$$

for some $c > 0$ and some $n \in \mathbb{N}$ (these functions are called polynomially bounded) are dense in the set of all functions holomorphic in G.

Here we refer to the topology of uniform convergence on compact subsets of G. As a consequence one can obtain the following results (without proof):

III.1.5 Theorem: (Pflug)

Let $G_1 \subset G_2$ be two domains and suppose that every function $f(z)$ holomorphic in G_1 and bounded by

$$|f(z)| \leq c\Delta^{-n}(z),$$

for some $c > 0$ and $n \in \mathbb{N}$, can be extended into G_2. Then every function holomorphic in G_1 can be extended into G_2. (Here $\Delta(z)$ is the generalized distance function belonging to the set G_1.)

This result is very useful since in several cases the envelope of holomorphy can be calculated by looking at tempered distributions. As an example of this statement one shows

III.1.6 Corollary

Let G be a connected domain $G \subset \mathbb{R}^n$ and let $T(G) = \{z \in \mathbb{C}^n; \operatorname{Im} z \in G\}$. Then

$$H(T(G)) = T(\operatorname{Co} G),$$

where $\operatorname{Co} G$ denotes the convex hull of G.

Proof: Without loss of generality one can assume $0 \in G$. If $|f(z)| \leq c\Delta^{-n}(z)$ then $f(x) \in \mathcal{S}'(\mathbb{R}^n)$ and hence $\mathcal{F}^{-1}(f) \in \mathcal{S}'(\mathbb{R}^n)$. Now the set of $y \in \mathbb{R}^n$ such that $e^{-(p,y)}\mathcal{F}^{-1}(f) \in \mathcal{S}'(\mathbb{R}^n)$ is convex (by Lemma II.1.2) and hence by Theorem II.1.3 $f(z)$ is analytic in $T(\operatorname{Co} G)$. By Theorem III.1.5 this is true for every $f(z)$ holomorphic in $T(G)$. On the other hand by Lemma III.1.3 (d), $T(\operatorname{Co} G)$ is a domain of holomorphy since it is convex and hence $H(T(G)) = T(\operatorname{Co} G)$. $\qquad\square$

Finally a result about functions of several complex variables has to be mentioned since it is often a tool in proofs.

III.1.7 Proposition:

Let $G \subset \mathbb{C}^n$ be a domain and $H(G)$ its envelope of holomorphy. Let $f(z)$ be a function holomorphic in G and hence in $H(G)$ and assume that $f(z) \neq w$ for $z \in G$. Then one also has

$$f(z) \neq w \quad \text{for all} \quad z \in H(G).$$

Proof: $\left(f(z) - w\right)^{-1}$ is holomorphic in G by assumption and can therefore be extended into $H(G)$. This implies that $f(z) \neq w$ for all $z \in H(G)$. $\quad \Box$

III.2 The edge of the wedge theorem

In the theory of functions of one complex variable the following result is well known and easy to prove with help of the Cauchy integral formula. Let

$$D^+ = \{z; |z| < 1 \quad \text{and} \quad \text{Im } z > 0\},$$

and

$$D^- = \{z; |z| < 1 \quad \text{and} \quad \text{Im } z < 0\}.$$

Assume that there are two functions f^+, f^- holomorphic in D^+ and D^- respectively, and having continuous boundary values on the real axis which coincide for $|x| < 1$. Then there exists a function f holomorphic in the disk D with the property $f = f^+$ for $z \in D^+$ and $f = f^-$ for $z \in D^-$.

Generalizations of this theorem to functions of several complex variables are called edge of the wedge theorems. One should remark, however, that the real subspace of $\mathbb{C}^n$ no longer splits $\mathbb{C}^n$ into two parts. Therefore, it is necessary to replace the upper half-plane of $\mathbb{C}$ by some other object. Let $C \subset \mathbb{R}^n$ be a proper, convex, open cone. $T(C)$ denotes the tube

$$T(C) = \{z \in \mathbb{C}^n; \text{Im } z \in C\}.$$

Such tubes will replace the upper half-plane. The edge of the wedge theorem now reads:

III.2.1 Theorem: (Edge of the Wedge)

Denote by B the ball

$$B = \{z; \|z\| := (\sum |z_i|^2)^{1/2} < 1\}$$

and define $B_C^+ = B \cap T(C)$ and $B_C^- = B \cap T(-C)$. Assume that $f^+(z)$ and $f^-(z)$ are functions holomorphic in B_C^+ and B_C^- respectively with f^+ and f^- having continuous boundary values at real points $\|x\| < 1$, and that these *boundary values coincide. Then there exists a complex neighbourhood $\mathcal{N}$ of $\mathbb{R}^n \cap B$ and a function f holomorphic in $B_C^+ \cup B_C^- \cup \mathcal{N}$ such that*

$$f = f^+ \text{ on } B_C^+ \quad \text{and} \quad f = f^- \text{ on } B_C^-.$$

The proof of this theorem will be divided into several parts.

III.2.2 Lemma:

Denote by $K \subset \mathbb{R}^n$ the set $\{x; 0 < x_i < 1, i = 1...n\}$ and by $T(K)$ the tube with base K. Let $f^+(z)$ be a function holomorphic in $T(K)$ and $f^-(z)$ a function holomorphic in $T(-K) = -T(K)$. Let $\Delta(z)$ be as defined in the last section and assume that

$$|f^\pm(z)| \le c^\pm \Delta(z)^{n^\pm}$$

for suitable $c^\pm$ and $n^\pm$. Moreover, assume that f^+ and f^- have continuous boundary values $f^+(x)$ and $f^-(x)$ which coincide. Then there exists a function $f(z)$ holomorphic in $T(\mathrm{Co}\{K \cup -K\})$ with $f(z) = f^+(z)$ for $z \in T(K)$ and $f(z) = f^-(z)$ for $z \in T(-K)$.

Proof: By Theorem II.1.5 $f^+(z)$ and $f^-(z)$ both have boundary values in the sense of tempered distributions. Hence $f^+(x)$ and $f^-(x)$ are both tempered distributions and so $\mathcal{F}^{-1}f^+ = \mathcal{F}^{-1}f^- = g$ is also a tempered distribution. The set of y such that $e^{-(y,p)}g(p) \in \mathcal{S}'$ contains $\{0\}$, K and $-K$ by assumption, and hence by Lemma II.1.2 it follows that $e^{-(y,p)}g(p) \in \mathcal{S}'$ for $y \in \mathrm{Co}\{K \cup -K\}$. Taking the Fourier transform of g and using Theorem II.1.3 one obtains the statement of the Lemma. □

By a suitable transformation of the last result one obtains

III.2.3 Corollary:

Let $D^+ = \{z \in \mathbb{C}; |z| < 1 \text{ and } \mathrm{Im}\, z > 0\}$ and define

$$D_n^+ = D^+ \times D^+ \times ... \times D^+$$

(the n-th power of D^+) and $D_n^- = -D_n^+$. Let $f^\pm(z)$ be two functions holomorphic in D_n^+ and D_n^-, respectively. Assume f^+ and f^- are both bounded and have continuous boundary values $f^\pm(x)$. Then $f^+(x) = f^-(x)$ implies that there exists a complex neighbourhood $\mathcal{N}$ of the cube $|x_i| < 1$ and a function f holomorphic in $D_n^+ \cup D_n^- \cup \mathcal{N}$ which coincides with f^+ on D_n^+ and with f^- on D_n^-.

Proof: By the transformation

$$z_i = \frac{e^{w_i} - 1}{e^{w_i} + 1}$$

one obtains a bi-holomorphic transformation of the tube $T(\frac{\pi}{2}K)$ of Lemma III.2.2 onto D_n^+. Define $g^\pm(w) = f^\pm(z(w))$. Then $g^\pm(w)$ is bounded and hence Lemma III.2.2 is applicable. One obtains the result by transformation of the last Lemma. □

Proof of the Theorem: Let $y_1, ..., y_n$ be n linearly independent vectors belonging to the cone C. Let $x_0 \in B$ and choose $\alpha > 0$ such that $\|x_0\| + \alpha \sum \|y_i\| < 1$. Let $D_n(x_0) = \{x_0 + \alpha \sum \lambda_i y_i; |\lambda_i| < 1\}$, and let $D_n^\pm(x_0)$ be the same sets with $\lambda > 0$ and $\lambda < 0$ respectively. If now $f^\pm$ are functions holomorphic in $B^\pm$ which have continuous boundary values then $f^\pm$ are bounded on $D_n^\pm(x_0)$. Hence Corollary III.2.3 applies and we get a complex neighbourhood of x_0 into which f^+ and f^- can be extended and where these extensions coincide. Since x_0 was an arbitrary point the theorem is proved.□

Theorem III.2.1 can be generalized in two different directions, first

III.2.4 Corollary:

Let $G \subset \mathbb{R}^n$ and $G^+, G^- \subset \mathbb{C}^n$ be domains. Denote by B_α sets of the form $B_\alpha = \{z; \|x_\alpha - z\| < r_\alpha\}$. Assume there exists a family of balls $B_\alpha, \alpha \in I$ and a family of convex cones $C_\alpha, \alpha \in I$ subject to the following conditions:
(a) $B_\alpha \cap \mathbb{R}^n$ is a locally finite covering of G,
(b) $B_\alpha \cap B_\beta \neq \emptyset$ implies $C_\alpha \cap C_\beta \neq \emptyset$,
(c) $B_\alpha \cap T(C_\alpha) \subset G^+$,
(d) $B_\alpha \cap T(-C_\alpha) \subset G^-$.
Let f^+ and f^- be two functions holomorphic in G^+ and G^- respectively, and assume that f^+ and f^- both have continuous boundary values in G. If these boundary values coincide then there exists a complex neighbourhood $\mathcal{N}$ of G and a function f holomorphic in $G^+ \cup G^- \cup \mathcal{N}$ which coincides on G^+ with f^+ and on G^- with f^-.

Proof: Apply III.2.1 to $G^+ \cap B_\alpha$, $G^- \cap B_\alpha$, and $G \cap B_\alpha$. □

Up to now it has always been assumed that the two functions f^+ and f^- have continuous boundary values. However, this is not necessary. One may replace continuous boundary values by boundary values in the sense of distributions. The corresponding statement will be made only for Theorem III.2.1, although it is also true for more general domains as in Corollary III.2.4.

III.2.5 Corollary:

Make the same geometric assumptions as in Theorem III.2.1. Let f^+ and f^- be two functions holomorphic in B_C^+ and B_C^- respectively. Assume f^+ and f^- both have boundary values in the sense of distributions on $B \cap \mathbb{R}^n$ and assume that these boundary values coincide in the sense of distributions. Then one has the same conclusion as in Theorem III.2.1.

Proof: Denote by B_r the ball of radius r, $B_r = \{z; \|z\| < r\}$ and identify B with B_1. Let $g \in \mathcal{D}$ with supp $g \subset B_\epsilon \cap \mathbb{R}^n$. Then $f^+ \star g$ and $f^- \star g$ (where $\star$ denotes the convolution) fulfil the assumptions of Theorem III.2.1 for the ball $B_{1-\epsilon}$. So there exists a function $f_g(z)$ holomorphic in $B_{1-\epsilon}^+ \cup B_{1-\epsilon}^- \cup \mathcal{N}(B_{1-\epsilon} \cap \mathbb{R}^n)$ which extends $f^+ \star g$ and $f^- \star g$. In order to obtain the result one has to remove the test function g. Let $r < 1$. Then f^+ and f^- are distributions of finite order on $B_r \cap \mathbb{R}^n$. Hence one obtains the estimate $|(f^\pm \star g)(z)| \leq \|g\|_m$ for $z \in B_{r-\epsilon}^\pm$ and some suitable norm $\|.\|_m$ on $\mathcal{D}$. We call the analytic continuation of the common function $(f \star g)(z) = f_g(z)$. Then it has an analytic continuation into the envelope of holomorphy $H(B_{r-\epsilon}^+ \cup B_{r-\epsilon}^- \cup \mathcal{N}(B_{r-\epsilon} \cap \mathbb{R}^n))$. This implies by Proposition III.1.7 the same estimate for $f_g(z)$, i.e. $|f_g(z)| \leq c\|g\|_m$ for $z \in H(B_{r-\epsilon}^+ \cup B_{r-\epsilon}^- \cup \mathcal{N}(B_{r-\epsilon} \cap \mathbb{R}^n))$. Call this domain Γ. Then we have an equi-continuous set of distributions $F_g(z, .)$ for $z \in \Gamma$. But on $B_{r-\epsilon}^\pm$ one has $F_g(z, x') = F_g(z - x')$ and this relation must be true on Γ by analytic continuation. Now $F_g(z)$ is a distribution which satisfies the Laplace equation in x_i, y_i and it is C^∞ in Γ and therefore also analytic. But for $z \in \mathbb{R}^n$ it has the form $(f \star g)(x)$ and by analytic continuation we conclude that there exists $F(z)$ with $F_g(z) = (F \star g)(z)$. This $F(z)$ extends f^+ and f^- in Γ. Now taking $r \to 1$ and $\epsilon \to 0$ one obtains a function $F(z)$ which is analytic in $B^+ \cup B^- \cup \mathcal{N}(B \cap \mathbb{R}^n)$. $\qquad \square$

III.3 The double cone theorem

The theory of several complex variables is a powerful tool in quantum field theory, in particular for enlarging domains of analyticity. However, of interest in physics are usually the real points belonging to the envelope of holomorphy. Therefore one is interested in techniques for finding some of these real points. A result of this type is the double cone theorem. It will be represented in a very general form, which will be needed later.

III.3.1 Notation:

(1) *Let C be a convex open cone in $\mathbb{R}^n$ and $a, b \in \mathbb{R}^n$ such that $b - a \in C$. Let $D_{a,b}$ denote the order interval*

$$D_{a,b} = \{a + C\} \cap \{b - C\}.$$

For $t \in C$, $D_{-t,t}$ will be denoted, for brevity, by D_t.
(2) *Let $C_1 = \{y \in C; \|y\| = 1\}$ and let $r(x, y)$ be a function on $D_{a,b} \times C_1$ with $r(x, y) > 0$ which is lower semi-continuous.*
(3) *Define:*

$$\Delta^+ = \{z = x + iy \in \mathbb{C}^n; x \in D_{a,b}, \ y \in C, \text{ and } \|y\| < r(x, \frac{y}{\|y\|})\}$$

and

$$\Delta^- = \{z = x + iy \in \mathbb{C}^n; x \in D_{a,b}, \ -y \in C, \text{ and } \|y\| < r(x, \frac{y}{\|y\|})\}.$$

Since r is lower semi-continuous it follows that on every compact set this function has a minimum which is different from zero.

III.3.2 Theorem: (Double Cone Theorem)

Let G be a domain of $\mathbb{R}^n$, and let $\mathcal{N}(G)$ be some complex neighbourhood of G. Let $\Gamma = \Delta^+ \cup \Delta^- \cup \mathcal{N}(G)$ and $H(\Gamma)$ its envelope of holomorphy. Assume $c, d \in G$ such that $d - c \in C$ and $c + \lambda(d - c) \in G$ for $0 \le \lambda \le 1$. Then

$$D_{c,d} \subset H(\Gamma) \cap \mathbb{R}^n.$$

See Fig 1.a.

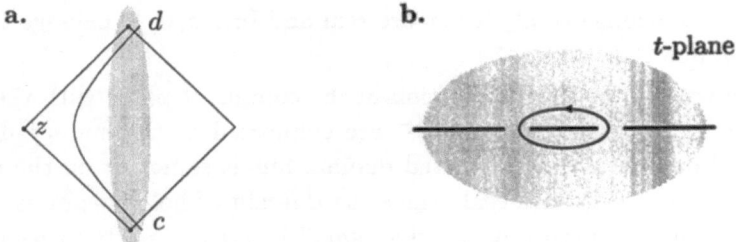

Fig. 1. Enlargement of the coincidence domain for the double cone theorem.
a. The shaded domain is the given coincidence domain. c and d are two points which are timelike to each other. The double cone is the enlargement obtained by the hyperboloids $C(\underline{\mu}, r, t)$ with center at $z = c + \sum \mu_i e_i$.
b. Analyticity domain in the t–plane and the curve for the Cauchy integral.

Proof: This theorem is proved if it can be demonstrated for a family of convex open subcones $C_\alpha \subset C$ such that to every direction in C there exists a subcone C_α containing this direction. To this end let $e_1, ..., e_n$ be n linearly independent vectors belonging to $\overline{C}$ such that $\sum e_i = d - c$. Then the subcone C_α in question is the cone $\{\sum \lambda_i e_i; \lambda_i > 0\}$ and the corresponding order interval is

$$D_{c,d}^{C_\alpha} = \{c + \sum \lambda_i e_i; 0 < \lambda_i < 1\}.$$

Now the curve $C(\underline{\mu}, r, t)$ given by the equation

$$x(\underline{\mu}, r, t) = c + \sum \mu_i e_i + tr \sum (1 - \mu_i)e_i - \frac{r}{t} \sum \mu_i e_i$$

defines a hyperboloid in the two-plane spanned by the three points c, d, and $c + \sum \mu_i e_i$. This hyperboloid has its center of mass at $c + \sum \mu_i e_i$ and has the asymptotes $c + \sum \mu_i e_i - \alpha \sum \mu_i e_i$ and $c + \sum \mu_i e_i + \alpha \sum (1 - \mu_i) e_i$ respectively, where $\alpha \in \mathbb{R}$.

We now choose $0 < \mu_i < 1$, $0 < r < \frac{1}{2}$ and restrict t to the interval:

$$\frac{1}{2r} - \sqrt{\frac{1}{4r^2} - 1} \le t \le \frac{1}{2r} + \sqrt{\frac{1}{4r^2} - 1}.$$

In this interval the function $\mu_i + tr(1 - \mu_i) - \frac{r}{t}\mu_i$ is monotone increasing in t and takes the values $\frac{1}{2} - \sqrt{\frac{1}{4} - r^2}$ and $\frac{1}{2} + \sqrt{\frac{1}{4} - r^2}$ at the end–points. This shows that in this interval $\mathcal{C}(\mu, r, t)$ belongs to $D_{c,d}^{C_\alpha}$ and that the end–points lie on the straight line connecting c with d.

If we vary μ then the curves $\mathcal{C}(\mu, r, t)$ cover all of $D_{c,d}^{C_\alpha}$. Moreover, $\operatorname{Im} x(\mu, r, t)$ belongs to C_α if μ, r are real and $\operatorname{Im} t > 0$. It belongs to $-C_\alpha$ if $\operatorname{Im} t < 0$. See Fig. 1.b.

Now we look at the intersection of the complex t-plane with the domain of holomorphy. Since Δ^+ and Δ^- are connected at the end–points of the t-interval one obtains a connected domain minus some cut on the real axis. This holds when μ, r are real. Since the domain of holomorphy is open one obtains a similar picture if one gives small imaginary parts to μ and r, e.g. a ring–shaped domain. Moreover, in the neighbourhood of $r = \frac{1}{2}$ the cut in the domain disappears. Now let $f(z)$ be holomorphic in $\Delta^+ \cup \Delta^- \cup \mathcal{N}(G)$ and define $g(\mu, r, t) = f(x(\mu, r, t))$. Then g can be extended by the Cauchy formula

$$G(\mu, r, t) = \frac{1}{2\pi i} \oint \frac{g(\mu, r, \tau)}{\tau - t} d\tau.$$

This function is analytic in a domain without a cut and since the cut disappears in a neighbourhood of $r = \frac{1}{2}$ one has $G(\mu, r, t) = g(\mu, r, t)$ for r close to $\frac{1}{2}$. Hence G is an extension of g. Since g depends only on the combination $x(\mu, r, t)$ this is also true for G. Hence F defined by $G(\mu, r, t) = F(x(\mu, r, t))$ defines an extension of f. By the edge of the wedge theorem we have analyticity in a full neighbourhood of G. Therefore we can deform the above Cauchy integral and obtain for $F(x(\mu, r, t)) = G(\mu, r, t)$ analyticity in a full neighbourhood of $D_{cd}^{C_\alpha}$. This function defines an extension of f. □

III.4 The Jost-Lehmann-Dyson representation

Let V^+ denote the open forward light cone

$$V^+ = \{x \in \mathbb{R}^n; x = (x_0, x_1, ..., x_{n-1}), \text{ with } x_0 > 0, \text{ and } x^2 > 0\}$$

where $x^2 = x_0^2 - \sum_1^{n-1} x_i^2$.

In applications of the theory of analytic functions to quantum field theory one often deals with the following situation. Let $f^+(z)$ and $f^-(z)$ be two analytic functions holomorphic in $T(V^+)$ and $T(-V^+)$ respectively, and let $G \subset \mathbb{R}^n$ be a domain (which does not have to be connected). Assume f^+ and f^- have boundary values for $y \to 0$, as functions or as distributions, and further assume that these boundary values coincide for $x \in G$. Then by the edge of the wedge theorem III.2.1 there exists a complex neighbourhood $\mathcal{N}(G)$ of G and a function f holomorphic in $T(V^+) \cup T(-V^+) \cup \mathcal{N}(G)$ such that f^+ and f^- are two different representations of f. Since f is holomorphic in $T(V^+) \cup T(-V^+) \cup \mathcal{N}(G)$ it is also holomorphic in the envelope of holomorphy of this set. This envelope will be denoted by $H(G)$.

For computing $H(G)$ one can use the theorem of Pflug III.1.5 and calculate the analytic continuation only for functions which are "polynomially" bounded. However, such functions have boundary values in $S'(\mathbb{R}^d)$ (Theorem II.1.5). These are f^+ and f^- respectively, which are Fourier transforms of tempered distributions having supports in $\overline{V}^+$ and $-\overline{V}^+$ respectively (Theorem II.1.7). Such distributions have been investigated extensively.

III.4.1 Lemma:

Let f be a tempered distribution such that supp $\mathcal{F}^{-1}f \subset \overline{V}^+ \cup -\overline{V}^+$ where $\mathcal{F}$ denotes the Fourier operator. Then

$$f \in C^\infty(\mathbf{x}, S'(x_0))$$

namely, f is a C^∞-function in the variables $x_1, ..., x_{n-1}$ with values in the space of tempered distribution in the time direction.

Proof: Since on $\overline{V}^+$ and $-\overline{V}^+$ one has $|x_0| \geq \|\mathbf{x}\|$ it follows that there exists a C^∞-function $b(x) = b(x_0, \mathbf{x})$ which is bounded and has bounded derivatives with the property:

$$(1 + x_0^2 + \mathbf{x}^2) = b(x)(1 + 2x_0^2), \quad x \in \overline{V}^+ \cup -\overline{V}^+.$$

Now let $f(x_0, \mathbf{x})$ be a tempered distribution with support in $\overline{V}^+ \cup -\overline{V}^+$. Then by the representation theorem for tempered distributions there exists a bounded distribution $F(x_0, \mathbf{x})$ and an $m \in \mathbb{N}$ with

$$f(x_0, \mathbf{x}) = (1 + x_0^2 + \mathbf{x}^2)^m F(x_0, \mathbf{x}).$$

Now take $\varphi(x_0) \in \mathcal{S}(\mathbb{R})$. Then for every $n \in \mathbb{N}$

$$(1 + x_0^2 + \mathbf{x}^2)^n f(x_0, \mathbf{x}) \varphi(x_0) = (1 + x_0^2 + \mathbf{x}^2)^{n+m} F(x_0, \mathbf{x}) \varphi(x_0)$$
$$= b(x)^{n+m} F(x_0, \mathbf{x})(1 + 2x_0^2)^{n+m} \varphi(x_0)$$

is a bounded distribution since $(1+2x_0^2)^{n+m} \varphi(x_0)$ is in $\mathcal{S}(\mathbb{R})$. But this shows that $f(x_0, \mathbf{x}) \varphi(x_0)$ is in the space $\mathcal{O}_c$ of Schwartz. Moreover, this expression depends continuously on $\varphi(x_0)$.

This argument remains true if we replace x by p and f by $\mathcal{F}^{-1}f$. Going back to the configuration space by Fourier transformation one obtains that $(f, \varphi(x_0)) \in \mathcal{O}_m$ i.e., in the space of tempered C^∞-functions. Clearly then $(f, \varphi(x_0))(\mathbf{x})$ is a C^∞-function for every $\varphi \in \mathcal{S}$ depending continuously on φ. But this is equivalent to the statement of the Lemma. $\quad\square$

This last Lemma has been proved in order to prepare the following

III.4.2 Theorem:

There is a one to one correspondence of the following two sets:
(1) *Tempered distributions f on $\mathbb{R}^n$ with the property:*

$$\operatorname{supp} \mathcal{F}^{-1}f \subset \overline{V}^+ \cup -\overline{V}^+.$$

(2) *Tempered distributions $F(x_0, x_1, ..., x_n)$ on $\mathbb{R}^{n+1}$ with the properties:*
(a) $F(x_0, x_1, ..., x_n)$ *is even in the last variable x_n.*
(b) $F(x_0, x_1, ..., x_n)$ *satisfies the wave equation*

$$\left(-\frac{\partial^2}{\partial x_0^2} + \sum_1^n \frac{\partial^2}{\partial x_i^2}\right) F(x_0, x_1, ..., x_n) = 0.$$

The correspondence is given by the formula

$$f(x_0, ..., x_{n-1}) = F(x_0, ..., x_{n-1}, 0)$$

or by

$$F(x_0, x_1, ..., x_{n-1}, x_n) = \mathcal{F}_{\mathbb{R}^n}\{(\mathcal{F}^{-1}f)(p) \cos x_n \sqrt{p^2}\}.$$

Proof: Assume f is given with $\operatorname{supp} \mathcal{F}^{-1}f \subset \{p; p^2 \geq 0\}$. Since the function $\cos \lambda$ depends only on λ^2 it follows that $\cos x_n \sqrt{p^2}$ is a C^∞-function in all variables. Moreover, on the set $\{p; p^2 \geq 0\}$ it is polynomially bounded together with all its derivatives. This shows that $(\mathcal{F}^{-1})(p) \cos x_n \sqrt{p^2}$ is a well-defined tempered distribution in all variables. Moreover, it is even in x_n. Differentiating the cosine twice one obtains

$$\left(p^2 + \frac{\partial^2}{\partial x_n^2}\right)(\mathcal{F}^{-1}f)(p) \cos x_n \sqrt{p^2} = 0.$$

Now define $F(x_0, ..., x_{n-1}, x_n)$ as the partial Fourier transformation with respect to the variables p, hence

$$\mathcal{F}_{\mathbb{R}^n}\{(\mathcal{F}^{-1}f)(p)\cos x_n \sqrt{p^2}\}$$

satisfies the wave equation

$$\left(-\frac{\partial^2}{\partial x_0^2} + \sum_1^n \frac{\partial^2}{\partial x_i^2}\right)F = 0.$$

Since the Fourier transform of F has its support on $\{(p, p_n); p^2 - p_n^2 = 0\}$ one can apply Lemma III.4.1 and put $x_n = 0$. Since the defining formula for F is continuous in x_n one is allowed to put $x_n = 0$ in this expression. One obtains $F(x_0, x_1, ..., x_{n-1}, 0) = f(x_0, ..., x_{n-1})$.

On the other hand assume that $F(x_0, x_1, ..., x_{n-1}, x_n)$ is a function which satisfies the wave equation, and define $f(x_0, ..., x_{n-1}) = F(x_0, x_1, ..., x_{n-1}, 0)$. Then one finds

$$(\mathcal{F}_{\mathbb{R}^n}^{-1}f)(p_0, ..., p_{n-1}) = \int (\mathcal{F}_{\mathbb{R}^{n+1}}^{-1}F)(p_0, ..., p_{n-1}, p_n)\, dp_n.$$

This expression is well-defined since $\mathcal{F}_{\mathbb{R}^{n+1}}^{-1}F$ has compact support in p_n for every $p_0, ..., p_{n-1} \in \mathbb{R}^n$. Since $\mathcal{F}F$ has its support on $\{(p, p_n); p^2 - p_n^2 = 0\}$ one finds supp $\mathcal{F}_{\mathbb{R}^n}^{-1}f \subset \overline{V}^+ \cup -\overline{V}^+$. So it remains to show that the restriction to the subspace $x_n = 0$ is one to one for such F which are even in the last variable. Assume F_1, F_2 are two solutions of the wave equation which are even in x_n and whose restrictions to $x_n = 0$ coincide. Then we have to show that $F_1 = F_2$. Let $F = F_1 - F_2$. Then $F|_{x_n=0} = 0$. F is even in the last variable which implies $\frac{\partial F}{\partial x_n}|_{x_n=0} = 0$, and hence by the wave equation $D^i F|_{x_n=0} = 0$. It remains to show that from this one can conclude that $F = 0$. This will be done in a separate lemma.

III.4.3 Lemma:

Let $F \in S'(\mathbb{R}^{n+1})$ be a solution of the wave equation and assume that $D^i F(x_0, 0) = 0$ for $|x_0| < 1$ for all multi-indices i. Then one has

$$F(x_0, \mathbf{x}) = 0 \quad \text{for} \quad |x_0| + \|\mathbf{x}\| < 1.$$

Proof: By Lemma III.4.1 the quantities appearing in the lemma are well-defined. Convoluting with a test function in x_0 having support in $|x_0| < \frac{\epsilon}{2}$ one can restrict attention to C^∞-functions. If the result is true for such functions one obtains the general result by letting the test function tend to the δ-function.

First assume that $G(y_1, ..., y_n, x_1, ..., x_n)$ is a C^∞-function of $2n$ variables satisfying the equation

$$\sum_{i=1}^{n} \frac{\partial^2}{\partial y_i^2} G = \sum_{i=1}^{n} \frac{\partial^2}{\partial x_i^2} G.$$

Let $\widetilde{G}(p, q)$ denote the inverse Fourier transform of G. Then $\widetilde{G}(p, q) = 0$ unless $p^2 = q^2$. Hence one obtains:

$$\int \widetilde{f}(p^2)\widetilde{G}(p, q) \, dp \, dq = \int \widetilde{f}(q^2)\widetilde{G}(p, q) \, dp \, dq$$

for any C^∞-function $\widetilde{f}$ belonging to $\mathcal{O}_m$ which is the space of functions that, together with all their derivatives, are polynomially bounded. Taking the Fourier transformation of this equation one obtains

$$\int f(x^2)G(0, x) \, dx = \int f(y^2)G(y, 0) \, dy$$

provided f is a strongly decreasing distribution. One obtains the connection with the above problem by assuming that G depends only on y_1 and not on $y_2, ..., y_n$. Denoting y_1 by x_0 and identifying G with $D^i F$ one obtains

$$\int f(x^2)D^i F(0, x) \, dx = 0$$

provided supp $f \subset \{x^2 < 1 - \epsilon\}$. By partial integration one finds

$$\int \{D^i f(x^2)\} F(0, x) \, dx = 0$$

for any multi-index i and any $f(x^2)$ with the above support restriction. Since $\frac{\partial x^2}{\partial x_i} = 2x_i$ it follows that the distributions of the form $D^i f(x^2)$ are dense in the set of distributions with support in $\{x^2 < 1 - \epsilon\}$. From this one concludes that $F(0, x) = 0$ for $x^2 < 1 - \epsilon$. Taking now the limit $\epsilon \to 0$ and changing the origin of the x_0-axis one has $F(x_0, x) = 0$ for $|x_0| + \|x\| < 1$. □

Proof of the theorem (continued): We had concluded $D^i F(x_0, 0) = 0$ for every multi-index i. Hence by the last lemma we obtain $F(x_0, x) = 0$. Thus the map $F(x_0, x_1, ..., x_{n-1}, x_n) \to F(x_0, x_1, ..., x_{n-1}, 0)$ is unique if F satisfies the wave equation and is even in the last variable. □

To simplify the writing let us introduce the following

III.4.4 Notation:

(a) *Let $V^+, \widehat{V}^+$ be the forward light-cones in $\mathbb{R}^n$ and $\mathbb{R}^{n+1}$ respectively.*
(b) *A domain G in $\mathbb{R}^n$ is called order convex if it fulfils the identity*

$$G = \{G + V^+\} \cap \{G - V^+\}.$$

(c) *Let G be an order convex set in $\mathbb{R}^n$. We denote by $\widehat{G}$ the following set in $\mathbb{R}^{n+1}$:*

$$\widehat{G} = \{G + \widehat{V}^+\} \cap \{G - \widehat{V}^+\}.$$

(d) *Let G be a set in $\mathbb{R}^n$. We set*

$$G' = \{y; = (x-y)^2 < 0 \quad \text{for all} \quad x \in G\}.$$

(e) *A domain G is called a Jost-Lehmann-Dyson domain (in brief a J.L.D.-domain) if*

    α. *G is order convex and*
    β. *there is a spacelike hypersurface Σ,*

$$\Sigma \subset \overline{\widehat{G}} \cup \widehat{G'} \subset \mathbb{R}^{n+1},$$

which can serve as surface for the initial-value problem for the whole of $\mathbb{R}^n$.

(f) *$D(x_0, ..., x_n)$ denotes the Green function of the wave equation in $n+1$ dimensions, i.e. D is a solution of the wave equation which is odd in x_0 such that $D(0, \mathbf{x}) = 0$ and $\frac{\partial}{\partial x_0} D(0, \mathbf{x}) = \delta(\mathbf{x})$ which implies ($\epsilon(\lambda) = \text{sign}(\lambda)$)*

$$D(x_0, \mathbf{x}) = \mathcal{F}\{\epsilon(p_0)\delta(p_0^2 - (\mathbf{p})^2)\}.$$

Remark:
At this point it should be remarked that the set $\widehat{G'}$ is automatically closed when G is open. This can be seen as follows: Let x_n be a convergent sequence in $\widehat{G'}$ with limit x. From $(x_n - y)^2 < 0$ for all $y \in G$ one has $(x-y)^2 \leq 0$ for all $y \in G$. Assume there is $y \in G$ with $(x-y)^2 = 0$. Since G is open there exists an $\epsilon > 0$ such that $y + B_\epsilon \subset G$. Now let n_0 be such that $x - x_n \in B_{\epsilon/2}$ for $n > n_0$. Then one has $y - (x - x_n) \in G$ and $(x_n - y + (x - x_n))^2 = (x-y)^2 = 0$ contradicting the construction of $\widehat{G'}$. Hence $\widehat{G'}$ is closed. With the above notation one obtains the following important result:

III.4.5 Theorem: (Dyson)

(a) *Let G be a J.L.D. domain in $\mathbb{R}^n$ and denote by $\mathcal{K}(G)$ the set of distributions $f \in S'(\mathbb{R}^n)$ with*

(α) supp $\mathcal{F}^{-1}f \subset \overline{V}^+ \cup -\overline{V}^+$.
(β) $f(x) = 0$ for $x \in G$.

($\mathcal{K}$ stands for commutator function).

(b) *Denote by $\mathcal{D}y(G)$ the set of distributions $\psi \in S'(\mathbb{R}^{n+1})$ with*

(α) $\psi = \psi(x_0, \mathbf{x})$ *is strongly decreasing in x_0.*
(β) supp $\psi \subset \widehat{G'}$.
(γ) $\psi(x_0, ..., x_n)$ *is even in x_n.*

Then the linear map from $\mathcal{D}y(G)$ to $\mathcal{K}(G)$, given by the formula

$$\Delta(\psi) = (D \star \psi)(x_0, ..., x_{n-1}, x_n)|_{x_n=0},$$

is surjective.

Proof: The Green function D vanishes for $x^2 < 0$ which means it has, for fixed x_0, compact support in **x**. Hence $D \star \psi$ is well-defined if ψ is strongly decreasing in x_0. If $\psi \in \mathcal{D}y(G)$ then it follows from the support properties of D that $D \star \psi$ vanishes in $\widehat{G}'' \supset \widehat{G}$ and consequently its restriction to $x_n = 0$ vanishes in G and hence $\Delta(\psi) \in \mathcal{K}(G)$. It remains to show that the map Δ is surjective. Let $f \in \mathcal{K}(G)$. By Theorem III.4.2 there exists an element $F \in \mathcal{S}'(\mathbb{R}^{n+1})$ which is even in x_n and which fulfils the wave equation such that $F(x_0, ..., x_{n-1}, 0) = f$. Applying Lemma III.4.3 one finds that F vanishes in $\widehat{G}$ since the order convex hull coincides with the union of the order intervals between all pairs of G. On the other hand since F satisfies the wave equation one can solve the Cauchy initial value problem with respect to any spacelike hypersurface Σ and obtain ($x \in \mathbb{R}^{n+1}$)

$$F(x) = \int_{\Sigma} \sum_{\mu} \left[D(x-y) \frac{\partial F(y)}{\partial y^{\mu}} - \frac{\partial D(x-y)}{\partial y^{\mu}} F(y) \right] d\sigma^{\mu}(y).$$

Notice that G was order convex by assumption. Consequently $\widehat{G}$ is also order convex. In this situation $\overline{\overline{\widehat{G}}}$ and $\widehat{G}'$ intersect in a $(n+1) - 2$ dimensional manifold. Now One chooses a spacelike surface containing this intersection. Then the points in Σ belong either to $\widehat{G}$ or to $\widehat{G}'$. Since F vanishes on $\widehat{G}$ it follows that

$$F(x) = \int_{\Sigma \cap \widehat{G}'} \sum_{\mu} \left[D(x-y) \frac{\partial F(y)}{\partial y^{\mu}} - \frac{\partial D(x-y)}{\partial y^{\mu}} F(y) \right] d\sigma^{\mu}(y).$$

The integration over the manifold $\Sigma \cap \widehat{G}'$ can be viewed as integrating distributions having support on $\Sigma \cap \widehat{G}'$. So one obtains

$$F(x) = \int_{\widehat{G}'} \left[D(x-y)\psi^1(y) - \sum_{\mu} \frac{\partial D(x-y)}{\partial y^{\mu}} \psi^2_{\mu} \right] dy.$$

Integration by parts gives

$$F(x) = \int_{\widehat{G}'} D(x-y)\psi(y) \, dy$$

with

$$\psi = \psi^1 + \sum_\mu \frac{\partial}{\partial y^\mu} \psi^2_\mu.$$

□

Remark:
These formal manipulations are justified if Σ is a smooth manifold. For more general situations one first has to convolute the initial function f by test functions φ with compact support in order to obtain smooth functions. Having done this one takes at the end the limit with φ tending to the δ-function.

As a consequence of the last theorem one obtains:

III.4.6 Corollary:

With the notation of the last theorem every distribution $f \in \mathcal{K}(G)$ vanishes in $\widetilde{G} = \widehat{G}'' \cap \{x; x_n = 0\}$, or in other words,

$$\mathcal{K}(G) = \mathcal{K}(\widetilde{G}).$$

Proof: Suppose $f \in \mathcal{K}(G)$. Then there exists $\psi \in \mathcal{L}(G)$ with $f = \Delta(\psi)$. But $D \star \psi$ vanishes in $\widehat{G}''$ and hence f vanishes in $\widehat{G}'' \cap \{x; x_n = 0\}$. □

The starting point was two functions f^+, f^- holomorphic in $T(V^+)$ and $T(-V^+)$ respectively. These functions are supposed to be polynomially bounded in these domains and it was assumed that the boundary values coincide on some coincidence domain G. If we assume that G is a J.L.D. domain then $f = f^+ - f^-$ is a function belonging to $\mathcal{K}(G)$ and thus has the representation described in Theorem III.4.5. In order to obtain f^+ one has to cut f in momentum space because $(\mathcal{F}^{-1}f^+)(p) = (\mathcal{F}^{-1}f)(p)$ on $\overline{V}^+ \setminus \{0\}$. By doing this cutting one obtains f^+ up to a distribution with support at the point zero. As a consequence:

III.4.7 Theorem:

Let G be a J.L.D. domain. Denote by $\mathcal{K}^+(G)$ the set of $f^+ \in \mathcal{S}'(\mathbb{R}^n)$ with

(α) $\operatorname{supp} \mathcal{F}^{-1}f^+ \subset \overline{V}^+$.

(β) *there exists $f^- \in \mathcal{S}'(\mathbb{R}^n)$ with $\operatorname{supp} \mathcal{F}^{-1}f^- \subset -\overline{V}^+$ and $f^+(x) = f^-(x)$ for $x \in G$.*

Then there exists a distribution $\psi \in \mathcal{D}y(G)$ with

$$f^+ = \Delta^+(\psi) + P(x)$$

where $P(x)$ is a polynomial and $\Delta^+(\psi)$ is defined by the formula

$$\Delta^+(\psi) = (D^+ \star \psi)(x_0, ..., x_{n-1}, 0)$$

where $D^+(x) = \mathcal{F}\{\Theta(p_0)\delta(p^2)\}$ is a positive energy solution of the wave equation in $\mathbb{R}^{n+1}$.

Proof: Notice $f^+ - f^-$ fulfils the conditions of Theorem III.4.5. Hence there exists $\psi \in \mathcal{D}y(G)$ with $f^+ - f^- = \Delta(\psi)$. Let $F = D \star \psi$. Then one obtains

$$\mathcal{F}^{-1}F = \epsilon(p_0)\delta(p^2)(\mathcal{F}^{-1}\psi)(p)$$

and hence

$$\mathcal{F}^{-1}(f^+ - f^-) = \int \mathrm{d}p_n \epsilon(p_0)\delta(p^2)(\mathcal{F}^{-1}\psi)(p).$$

Doing the cutting one finds

$$\mathcal{F}^{-1}f^+ = \int \mathrm{d}p_n \Theta(p_0)\delta(p^2)(\mathcal{F}^{-1}\psi)(p) + \text{``}\delta\text{''}$$

where "δ" is a distribution with support at zero. Taking the Fourier transform of this it becomes

$$f^+ = (D^+ \star \psi)(x_0, ..., x_{n-1}, 0) = \Delta^+(\psi) + P(x). \qquad \square$$

Let G be a J.L.D. domain. Then, according to the remark at the beginning of this section, one can compute $H(G) =$ envelope of holomorphy of $T(V^+) \cup T(-V^+) \cup \mathcal{N}(G)$ by determining the domain of analyticity of all functions of the form $\Delta^+(\psi)$ with $\psi \in \mathcal{D}y(G)$. To this end one has to know the function D^+. It is invariant under Lorentz transformations, hence one obtains $D^+(x) = g(x^2)$. Since it has an analytic extension into the forward tube $T(V^+)$, $g(x^2)$ must be a function analytic in the cut complex x^2-plane where the cut runs along the positive real axis. Now the function $\{(x+iy)^2\}^{-(n-2)/2}$ satisfies, for fixed y, the wave equation and has the same analyticity properties as $D^+(x + iy)$. So one concludes that $D^+(x)$ has a singularity on the manifold $(x + iy)^2 = 0$.

On the other hand G is contained in $\mathbb{R}^n$. Therefore $\hat{G}$ is symmetric around the subspace $x_n = 0$. Furthermore, with $(x_0, ..., x_{n-1}, x_n)$ also $(x_0, ..., x_{n-1}, \lambda x_n)$ belongs to $\hat{G}$ for $\|\lambda\| \leq 1$. This in turn implies that if $(x_0, ..., x_{n-1}, x_n) \in \hat{G}'$ then also $(x_0, ..., x_{n-1}, \mu x_n) \in \hat{G}'$ for all μ with $|\mu| \geq 1$. Therefore one obtains the following characterization of $H(G)$.

III.4.8 Theorem:

Denote by $h(u)$ for $u \in \mathbb{R}^{n+1}$ the complex manifold

$$h(u) = \{z \in \mathbb{C}^n; (z_0 - u_0)^2 - \sum_{1}^{n-1}(z_i - u_i)^2 - u_n^2 = 0\}.$$

Let G be a J.L.D. domain and let

$H(G)$ = envelope of holomorphy of $T(V^+) \cup T(-V^+) \cup \mathcal{N}(G)$

where $\mathcal{N}(G)$ is a complex neighbourhood of G obtained by the edge of the wedge theorem. Then

$$H(G) = \mathbb{C}^n \setminus \left(\cup \{h(u); u \in \widehat{G}'\} \right)^-$$

where the bar indicates the closure of the set.

Proof: Denote by $\widetilde{u}$ the vector $(u_0, ..., u_{n-1})$ so that $u = (\widetilde{u}, u_n)$. Then the function $\{(z - \widetilde{u})^2 - u_n^2\}^{-1}$ has a singularity on $h(u)$. Therefore, $\mathbb{C}^n \setminus \left(\cup \{h(u); u \in \widehat{G}'\} \right)^- = H$ is a domain of holomorphy. For z in $T(V^+)$ or in $T(-V^+)$ one has $(z - \widetilde{u})^2 - u_n^2 \neq 0$. Moreover, for $u \in \widehat{G}'$ and $x \in G$, one has $(x - \widetilde{u})^2 - u_n^2 \neq 0$, so that H contains $T(V^+), T(-V^+)$ and G, and hence $H \supset H(G)$. Following the arguments at the beginning of this section it remains to show that every $f^+ \in \mathcal{K}^+(G)$ has an extension into H. But from the representation Theorem III.4.7 of such functions and the structure of the function D^+ one knows that f can be extended at least into the points $\mathbb{C}^n \setminus \cup \{z; (z - \widetilde{u})^2 - u_n^2 = \rho, \rho \geq 0\}$. However, this set coincides with H according to the special structure of $\widehat{G}'$. So we get $H = H(G)$. $\qquad \square$

III.5 Some consequences of the Jost-Lehmann-Dyson representation

Not every coincidence domain of the edge of the wedge problem is a J.L.D. domain, which means that it is not possible to compute the envelope of holomorphy for every edge of the wedge problem. On the other hand every coincidence domain contains J.L.D. domains whose envelopes of holomorphy are known. One should use this knowledge in order to obtain parts of the envelope of holomorphy of the general edge of the wedge problem. For application to physics one is mostly interested in the real points of the envelope of holomorphy and therefore our attention will focus on these.

We will apply the double cone theorem, which is a local result. Consequently it also can be used if the cone, on which the double cone theorem is based, changes from point to point. We start with the tube based on the light–cone and use the Jost-Lehmann-Dyson-representation for some J.L.D. domain. By this we obtain locally an enlarged cone which one is able to compute because the boundary of the J.L.D. problem is known. However, in order to make the investigation transparent we will restrict ourselves to the case where G is the spacelike complement of the double cone D_t.

Recall that the solution of the edge of the wedge problem in $\mathbb{C}^n$ is closely related to the solution of the wave equation in $\mathbb{R}^{n+1}$. For simpler notation we denote in this section the last coordinate by m so that one has

$$h(u, m) = \{z \in \mathbb{C}^n; (z - u)^2 - m^2 = 0\}.$$

For computing the envelope of holomorphy we need the set $\widehat{G}'$ which is in our case the set $D_t \times \mathbb{R}$. For the investigation we introduce the

III.5.1 Notation:

Let $H(D'_t)$ be the envelope of holomorphy for the edge of the wedge problem for the coincidence domain D'_t. For $x \in \mathbb{R}^n$ put

$$I_x = \{y \in \mathbb{R}^n; x + iy \in H(D'_t)\}.$$

Our first aim is to characterize the sets I_x.

III.5.2 Lemma:

The set I_x has the following properties:
(i) $y \in I_x$ implies that there exist two neighbourhoods $\mathcal{U}_x$ of x and $\mathcal{U}_y$ of y such that

$$y' \in I_{x'},$$

provided $y' \in \mathcal{U}_y$ and $x' \in \mathcal{U}_x$.
(ii) $y \in I_x$ implies $-y \in I_x$.
(iii) $V^+ \subset I_x$.
(iv) If $x \in D_t$ and $y \in I_x$ then one has $y^2 > 0$.
(v) If $y \in I_x$, then it follows that

$$\lambda y \in I_x \quad \text{for} \quad \lambda \in \mathbb{R} \quad \text{and} \quad 0 < |\lambda| \le 1.$$

(vi) If $x \in H(D'_t)$, then I_x contains a neighbourhood of zero in $\mathbb{R}^n$.

Proof: (i) This is due to the fact that $H(D'_t)$ is open in $\mathbb{C}^n$.
(ii) This follows from invariance of $H(G)$ under complex conjugation.
(iii) By assumption the tube $T(V^+)$ belongs to $H(D'_t)$.
(iv) Since $D_t \subset \widehat{G}'$ one has $(x, 0) \in \widehat{G}'$. Hence $-y^2 - m^2 \neq 0$ for all $m \in \mathbb{R}$. This is only possible for $y^2 > 0$.
(v) If y is timelike then the statement is true since the interior of the forward and backward light–cone always belongs to I_x. Assume now $y^2 \le 0$ and $y \in I_x$. Then for each $u \in D_t$ at least one of the two conditions must be fulfilled: Either $(x - u)^2 - y^2 < 0$ or $(x - u, y) \neq 0$. Since $y^2 \le 0$ these relations remain true if we replace y by λy with $0 < |\lambda| \le 1$.
(vi) If $x \in H(D'_t)$, then $\{0\} \in I_x$ and (vi) follows from (i). $\qquad \square$

Since I_x has the property that if $y \in I_x$ then λy also belongs to I_x for $0 < \lambda \le 1$ we can characterize I_x as follows: Let $\|y\| = \{\sum_{i=0}^{n-1} y_i^2\}^{1/2}$ and define

$$E_x = \{\frac{y}{\|y\|}; y \in I_x, y \neq 0\}$$

and

$$\sigma_x(y) = \sup\{\lambda; y \in E_x, \text{ and } \lambda y \in I_x\}.$$

Then one has for $x \notin H(D'_t)$

$$I_x = \{\lambda y; y \in E_x \text{ and } 0 < \lambda < \sigma_x(y)\}.$$

If $x \in H(D'_t)$ then one has

$$I_x = \{\lambda y; y \in E_x \text{ and } 0 \leq \lambda < \sigma_x(y)\}.$$

In order to characterize I_x we have to characterize the set E_x and the function $\sigma_x(y)$. But since we intend to apply only the double cone theorem we only need the structure of the sets E_x. The situation will become easier if we replace the sets E_x by the sets $\tilde{C}_x$ which are defined as follows

III.5.3 Definition:

We denote by $\tilde{C}_x$ the sets

$$\tilde{C}_x = \bigcup_{\lambda > 0} \lambda E_x$$
$$= \{y \in \mathbb{R}^n \setminus \{0\}; \exists \lambda > 0 \text{ such that } \lambda y \in I_x, \}$$

The structure of these sets is given in a sequence of three lemmas. We start with

III.5.4 Lemma:

For $x \in D'_t$ one has

$$\tilde{C}_x \cup \{0\} = \mathbb{R}^n.$$

Proof: This follows from $D'_t \subset H(D'_t)$ and Lemma III.5.2 (vi). □

Next consider points belonging to D_t.

III.5.5 Lemma:

Let $x \in D_t$. Then we have

$$\tilde{C}_x = V^+ \cup \{-V^+\}.$$

Proof: If $x \in D_t$ then all hyperboloids $h(x, m), m \geq 0$ with center in x belong to the complement of $H(D'_t)$. But these contain all y with $y^2 \leq 0$. □

It remains to investigate the complement of $D_t \cup D_t'$. This consists of three parts, the points $(x, t) > 0$, $(x, t) < 0$ and $(x, t) = 0$. The latter part is the intersection of the boundaries of D_t and D_t'. Because of symmetry it is sufficient to study the elements in the part $(x, t) \geq 0$.

In order that a point y belong to I_x it is necessary that for every $u \in \overline{D}_t$ at least one of the two conditions be fulfilled: Either $(x - u)^2 - y^2 < 0$ or $(u - x, y) \neq 0$. Therefore y belongs to $\tilde{C}_x$ if for every $u \in \overline{D}_t$ at least one of the two conditions, either $(x - u)^2 < 0$ or $(u - x, y) \neq 0$, holds. If a point $u \in \overline{D}_t$ is not spacelike with respect to x then the second condition must be fulfilled. Therefore we introduce

III.5.6 Definition:

For $x \in \left(\{-t + \overline{V}^+\} \setminus D_t \right) \cap \left((x, t) \geq 0 \right)$ the set P_x is defined as follows:

$$P_x = \{(x - u); (x - u) \in V^+; u \in \overline{D}_t\}.$$

See Fig. 2.

P_x is the intersection of two convex sets and hence it is again convex. In addition, this set is compact. If $y \in I_x$ then we have $(x - u)^2 \geq 0$ for $(x - u) \in P_x$. Hence $(x - u, y)$ must have only one sign if $(x - u)$ varies over P_x.

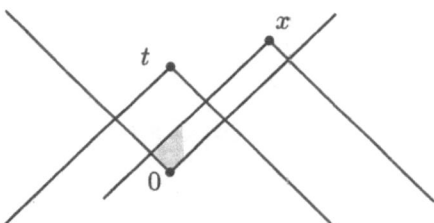

Fig. 2. The construction of the set P_x. The shaded set is $x - P_x$.

These remarks lead to

III.5.7 Lemma:

For $x \in \left(\{-t + \overline{V}^+\} \setminus D_t \right) \cap \left((x, t) \geq 0 \right)$ the set $\tilde{C}_x$ is the union of two convex cones C_x and $-C_x$ with

$$C_x = \{y \in \mathbb{R}^n; (u - x, y) > 0 \; \forall \; (u - x) \in P_x\}.$$

Proof: First note that for the points under consideration the set P_x is not empty. If $y \in \tilde{C}_x$ then for $u \in \overline{D}_t$ but $u \ni x - \overline{V}^+$ we have $(x-u)^2 < 0$. Hence for $(x - u) \in P_x$ we must have $(x - u, y) \neq 0$. Therefore, by compactness of P_x, we have $\inf |(x - u, y)| > 0$. Defining C_x as the set of those y for which the scalar product is positive we find the stated result. Since P_x is convex we see that C_x is a convex and open cone. $\qquad\square$

Now we want to define the cone C_x for arbitrary points.

III.5.8 Definition:

We define:

$$
C_x = \begin{cases}
\mathbb{R}^n, & \text{if} & x \in D'_t, \\
V^+, & \text{if} & x \in D_t, \\
\{y; (x - u, y) > 0, \\
\quad \forall\, (x - u) \in P_x\} & \text{if} & x \in \{-t + \overline{V}^+\} \setminus D_t,\ (x,t) \geq 0, \\
C_{(-x)} & \text{if} & x \in \{t - \overline{V}^+\} \setminus D_t,\ (x,t) \leq 0.
\end{cases}
$$

As a consequence of the characterization of the cone C_x we observe

III.5.9 Proposition:

With the exception of points belonging to the boundary of D'_t the cone C_x depends continuously on x.

Proof: If $x \in D'_t$ then $C_x = \mathbb{R}^n$ is constant. If $x \in D_t$ then $C_x = V^+$. If $x \in \{-t + V^+\} \setminus D_t$ then $C_x = \{y; (x - u, y) > 0 \quad \forall x - u \in P_x$. Since the intersection of $\overline{D}_t$ with $x - \overline{V}^+$ changes continuously with x provided x does not belong to the boundary of $-t + V^+$ one sees that P_x and hence also C_x changes continuously with x. By symmetry the argument also holds for points in $t - V^+$. This shows that we only have a discontinuity if we cross the boundary of D'_t. $\qquad\square$

III.6 Applications: A hole in the spectrum

The technique developed in the last section will now be applied to a special situation. The coincidence domain will consist of two parts. G_1 will be the spacelike complement of the double cone D_t and G will be a double cone $D_{a,b}$, $b - a \in V^+$ in the complement of D'_t. Because of the symmetry of the problem we can assume without loss of generality that $D_{a,b} \subset -t + V^+$. We shall compute the enlargement of G which is enforced by the domain of holomorphy $H(D'_t)$. This result will be used in section 4 of the next chapter. In order to obtain the desired result, we have to compute the cones C_x given by $H(G_1)$, and afterwards we have to apply the general version of the double cone theorem III.3.2 to the boundary of the domain G.

Before we can apply the results of the last section we need a formulation of the double cone theorem useful for our purpose.

III.6.1 Theorem:

Assume for the edge of the wedge problem that we have the coincidence-domain $D'_t \cup G$ with $G \subset \{-t+V^+\}$. Let x be a boundary-point of G. Assume that for every neighbourhood $\mathcal{U}_\epsilon$ the two sets $\mathcal{U}_\epsilon \cap G \cap C_x$ and $\mathcal{U}_\epsilon \cap G \cap -C_x$ are not empty. Then the envelope of holomorphy contains a neighbourhood of the point x.

Proof: From proposition III.5.9 we know that the cone C_x depends continuously on x. Hence there exists a neighbourhood $\mathcal{U}$ of x such that, with $C_\mathcal{U} = \{\cap_{x \in \mathcal{U}} C_x\}^\circ$, the two sets $\mathcal{U}_\epsilon \cap G \cap C_\mathcal{U}$ and $\mathcal{U}_\epsilon \cap G \cap -C_\mathcal{U}$ are not empty. This implies that we can find points u, v such that $u \in (x - C_\mathcal{U}) \cap \mathcal{U} \cap G$ and $v \in (x + C_\mathcal{U}) \cap \mathcal{U} \cap G$ and moreover, these two points are connected by a $C_\mathcal{U}$-timelike curve which lies completely in G. Applying the usual double cone theorem (Th. III.3.2) to this situation we see that x belongs to the envelope of holomorphy. □

More colloquially, this result can be formulated as follows: The boundary Σ of $H(G_1 \cup G) \cap \mathbb{R}^n$ must, at any point $x \in \Sigma$, be "spacelike" with respect to the local cone C_x.

For the formulation of the main result of this section we need some notations.

III.6.2 Definition:

With the assumptions of this section let
(i)
$$F = \{(u, m) \in D_t \times \mathbb{R}; h(u, m) \cap D_{a,b} = \emptyset\},$$

(ii)
$$\widetilde{G}_0 = \mathbb{R}^n \setminus \{h(u, m); (u, m) \in F\}^-$$

(iii) *and $\widetilde{G} =$ connected component of $\widetilde{G}_0$ containing $D_{a,b}$.*

Remark:

If we look at $\mathbb{R}^n \setminus \{h(u, m); (u, m) \in F\}^-$ then, depending on the size of $D_{a,b}$, this set has two or three components. One of them is D'_t for which we have used the J.L.D.-representation. The second component belongs to $-t+V^+$ and it contains $D_{a,b}$. This is the set we denoted by $\widetilde{G}$. If $D_{a,b}$ is large enough then there appears a third set which lies in $t - V^+$. It is not known whether this third set also appears in case one would be able to compute the envelope of holomorphy. We will only look at the set $\widetilde{G}$. Our result is:

III.6.3 Theorem:

Let $G_1 = D'_t$ and $G = D_{a,b} \subset \{-t + V^+\}$ then the envelope of holomorphy of the edge of the wedge problem with coincidence-domain $D'_t \cup D_{a,b}$ contains the real points $\widetilde{G}$.

Proof: If $G \subset D_t$ then $C_x = V^+$ and hence there is no enlargement by the double cone theorem. In this case it is in accordance with $\widetilde{G} = G$.

For the general case we remark first that an enlargement beyond $\widetilde{G}$ is impossible, since its boundary is given by a family of hyperboloids which do not enter G_1 and G.

In order to show that the statement of the theorem is correct, we will construct an interpolating family of domains G^λ, $0 < \lambda \leq 1$, with $G^\lambda = G$ for $\lambda < \lambda_0$, and $G^1 = \widetilde{G}$. This family will be increasing: $G^\lambda \subset G^{\lambda'}$ for $\lambda < \lambda'$ and it will depend continuously on λ:

$$G^\lambda = \bigcup_{\lambda' < \lambda} G^{\lambda'}$$

$$G^\lambda = \bigcap_{\lambda' > \lambda} G^{\lambda'}.$$

Now we will show that for $\lambda < 1$ the domain G^λ can be enlarged beyond every boundary-point which is not already a boundary-point of $\widetilde{G}$. Analytic continuation through every boundary-point which does not lie on the final boundary implies that we have analyticity in $G^{\lambda'}$ for some $\lambda' > \lambda$. From this we obtain by induction that $G^1 = \widetilde{G}$ belongs to the domain of holomorphy.

We define G^λ similarly to $\widetilde{G}$. We take more hyperboloids, namely those whose centers are situated in $D_{t/\lambda}$:

$$G^\lambda = \text{connected component of } \mathbb{R}^n \setminus \{h(u,m); u \in D_{t/\lambda}, h(u,m) \cap D_{a,b} = \emptyset\}.$$

Connected component means again the component containing $D_{a,b}$. Since $D_{t/\lambda}$ increases with decreasing λ it follows that G^λ increases with λ. Moreover, since t is timelike there exists λ_0 such that $b \in D_{t/\lambda_0}$, and hence $D_{a,b} \subset D_{t/\lambda_0}$. In this case one has $G^{\lambda_0} = G$.

If $x \in \partial G^\lambda$ then there exists $u \in D_{t/\lambda}$ and $m \in \mathbb{R}$ with $x \in h(u,m)$. By the definition of G^λ this hyperboloid must contain one of the two points, either a or b. This allows us to separate ∂G^λ into two parts, namely:

$$\partial^+ G^\lambda = \{x \in \partial G^\lambda; x \in h(u,m) \text{ and } b \in h(u,m)$$
$$\text{for a suitable } (u,m) \in D_{t/\lambda}\} \times \mathbb{R}\},$$
$$\partial^- G^\lambda = \{x \in \partial G^\lambda; x \in h(u,m) \text{ and } a \in h(u,m)$$
$$\text{for a suitable } (u,m) \in D_{t/\lambda}\} \times \mathbb{R}\}.$$

If $x \in \partial^+ G^\lambda$ then we can find the origin of the hyperboloid $h(u_0, m)$ passing through x and b. Since both points are on the same hyperboloid we have $(x - u_0)^2 = (b - u_0)^2 = m^2$. Suppose that δx and δu are small variations such that $\delta x \in -V^+$ and δu is such that $u_0 + \delta u \in D_{t/\lambda}$. Then we have $(x_0 + \delta x - u_0 - \delta u)^2 \neq (b - u_0 - \delta u)^2$. This equation implies $2(x - u_0, \delta x) - 2(x - b, \delta u) + (\delta x)^2 - 2(\delta x, \delta u) \neq 0$. Since $(x - u) \in V^+$ we conclude by the choice of the variations that $(x - b, \delta u) \geq 0$. This implies that u_0 is a supporting point of the set of all possible points u_0 with respect to the hyperplane perpendicular to $x - b$. This set in question is $D_{t/\lambda} \cap (b - V^+)$.

There are two cases: First, $x - b$ is lightlike, then b, x, u_0 lie on the same light-ray in $b - \partial V^+$. In this case u_0 is not unique, but we may choose it in $t^\perp$. Second, if $x - b$ is spacelike then the hyperplane perpendicular to $x - b$ contains a timelike vector which means that the supporting point u_0 is unique and it lies in $t^\perp$.

Let us denote by $s(x)$ the tangent vector at $h(u_0, m)$ which lies in the two-plane through the points b, x, and u_0. Then u_0 is a supporting point also for the hyperplane perpendicular to $s(x)$. This one sees as follows: In the first case because $(x - u_0)$ and $s(x)$ are both lightlike, and in the second case because the supporting hyperplane is not unique. This means if we choose the direction of $s(x)$ such that the angle between $s(x)$ and $x - b$ is small then we find that $(s(x), x - u) \geq 0$ for all $u \in D_{t/\lambda} \cap (b - V^+)$.

If x does not already belong to $\partial^+ \widetilde{G}$ then u_0 does not belong to D_t. In this case $(s(x), x - u) \geq c > 0$ for all $u \in D_t \cap (b - V^+)$ and hence $s(x) \in C_x$, and by the double cone theorem there exists an extension beyond the point x. With the same technique we also find that one can extend beyond the points of $\partial^- G^\lambda$ provided they do not belong to $\widetilde{G}$. With this technique we do not get an enlargement beyond the point lying on $\partial^+ G^\lambda \cap \partial^- G^\lambda$. Beyond these points we obtain the continuation by applying the double cone theorem directly. This is possible, since at such a point both tangent hyperplanes enter into the cone C_x provided x is not a final point. This proves the theorem. $\square$

III.7 Notes and remarks

(1) There exist many textbooks on the theory of functions of several complex variables. It is impossible to list all of them. A representative selection is Behnke and Thullen [[BT]], Bochner and Martin [[BM]], Vladimirov [[Vla]], Hörmander [[Hör]] and Gunning and Rossi [[GR]]. For a domain $G \subset \mathbb{C}^n$ to be a domain of holomorphy we have required that to every boundary-point z_0 one can find a function holomorphic in G which cannot be extended into a neighbourhood of z_0. From this one can deduce that there exists a function which has G as its domain of definition, or in other words this function cannot be further extended into any neighbourhood of any boundary-point. For constructing enlargements of a given domain the method of cutting "noses"

entering into the domain is extensively explained in [[BM]]. Modern text-books prefer the concept of holomorphic convexity introduced by Cartan und Thullen [CT] or the closely related concept of pseudo-convexity, see e.g. [Bre54,56]. These concepts are consequences of the result in Proposition III.1.7.

The density theorem for polynomially bounded functions is relatively modern and can be found in [Pf74]. This result is a final step in a long chain of density results. The conclusion by P. Pflug which can be found in [Pf82] is extremely useful, at least for situations appearing in physics, since one can make use of other branches of analysis for computing envelopes of holomorphy.

For the tube theorem (Corollary III.1.6) there exist a large number of different proofs. There is a proof due to Stein [Ste] showing that by "cutting-noses" also one can obtain this result. It might be worthwhile to mention summer school lectures on the theory of several complex variables which are designed to give an introduction for physicists. Examples are the lectures of Wightman [Wi60] and Epstein [Ep65].

(2) The "edge of the wedge" theorem was found by Bremermann, Oehme, and Taylor [BOT]. A simplified proof is due to F. Browder [Brow]. The situation treated in this section is called the opposite edge of the wedge problem. The general situation, where the two tubes in question are not opposite, has been treated by Epstein [Ep60].

(3) Every domain of holomorphy in $\mathbb{C}^n$ has to fulfil some convexity condition, which is called pseudo-convexity. If now such a domain of holomorphy has an intersection with $\mathbb{R}^n$ then one can expect that this intersection has also to fulfil some convexity criterion. The double cone theorem is of this nature. The first proof of this result was given by Vladimirov [Vl] and independently but a little later by the author [Bch61]. If the tubes which appear in the double cone theorem are based on the forward light-cone, then this tube is invariant under the map $z \to \frac{-z}{z^2}$. By this transformation the double cone can be mapped onto the forward light-cone. For the invariance-group of the forward tube see e.g. J. Bros [Bros].

(4) With the Jost-Lehmann-Dyson representation one has hit on an extremely fortunate situation. All attempts to generalize this method to other cones appearing in physics have been without sucess so far. The reason for this failure was the fact that it led to an overdetermined system of differential equation for which the Green's function method does not lead in the complete solution. The Jost-Lehmann-Dyson representation in the form presented here is due to Dyson [Dy]. Jost and Lehmann have solved a special case [JL] but with a different method. The Jost-Lehmann-Dyson representa-

tions deal only with a special family of functions. Therefore it was not clear at that time whether this representation leads to the envelope of holomorphy of the corresponding edge of the wedge problem. That this was indeed the case was proved by Bros, Messiah and Stora [BMS]. Nowadays we can conclude this directly from the result of Pflug [Pf74]. Lemma III.4.3 is called Asgeirsson's mean value theorem [Asg]. A detailed discussion of this result can be found in Courant and Hilbert [[CH]]. Theorem III.4.6 is also known by the name "reentrant nose theorem".

(5,6) The techniques presented in these two sections have been developed by the author. The first presentation of these results can be found in [Bch85].

Chapter IV

Locality Condition and the Spectrum of Translations

In this chapter we want to look at the interplay between the spectrum and the locality conditions. It is our aim to deduce some consequences for the translation group. We will mainly use the techniques described in the last chapter.

However, we first have to describe a method which connects the matrix elements of the translation group with expectation values of products of local observables. This is done in the first section. If there exists a one-parameter symmetry group and if the group representation fulfils the spectrum condition then, as we have seen in II.4, there exists a unique minimal representation of this group. In higher dimensions, however, such a result need not be true unless the algebraic structure of the C^*–dynamical system and the shape of the spectrum fit together. In section 2 we will show that the locality condition is such an exceptional case if the cone in question is the light cone.

That this unique minimal representation has remarkable features will be shown in the third, the fourth, and the sixth sections. In section 3 we show that the support of spectrum of the translations is always a set which is invariant under Lorentz transformations. This is also true if the Lorentz group is not a symmetry of the theory. This shows that the principle of maximal velocity for signals has far-reaching consequences which are in accordance with the usual particle interpretation of quantum field theory.

In the fifth section we deal with the following conjecture: If we have a representation of the algebra of local observables with spectrum condition, if the spectrum of the unique minimal representation starts at m_0, and if there is an upper gap in the spectrum between m_1 and m_2, then one must have $m_2 \leq 3m_0$. Although a proof for the general case is still missing the interpretation of this result has to be seen in connection with the theory of superselection sectors and the existence of anti-particles. In the fourth section we shall prove that, in the vacuum sector, the spectrum of the translation is an additive set.

The last section is reserved for examples and some simple general results. We give some applications of the result obtained so far, namely we show that the spectrum condition is a quantum mechanical condition which cannot be transcribed to classical systems. Furthermore we will show that it is impossible to associate operators to points.

With the results of chapter four it is also possible to show the independence of the locality axiom and the spectrum condition. But the corresponding examples will be given in notes and remarks.

IV.1   Commutators and the edge of the wedge problem

In this section we want to relate quantum field theory with the techniques developed in the last chapter.

Let $\{\mathcal{A}(O), \mathcal{A}, \mathbb{R}^d, \alpha\}$ be a local ring system and let $\{\pi, \mathcal{H}, U(a), \overline{V}^+\}$ be a representation of this system fulfilling the spectrum condition. Then $U(a) \in \pi(\mathcal{A})''$ and it has the representation

$$U(a) = \int_{\overline{V}+} e^{i(p,a)} dE(p).$$

We want to investigate certain expectation values of the group representation $U(a)$.

For simpler notation we write again x instead of $\pi(x)$. For $\psi \in \mathcal{H}$ and $x \in \mathcal{A}$ we define

$$F^+_{x,\psi}(a) = (\psi, x^* U(a) x \psi),$$
$$F^-_{x,\psi}(a) = (U(-a)\psi, x U(-a) x^* U(a)\psi) = (\psi, \alpha_a(x) x^* U(a)\psi)$$

and

$$F_{x,\psi}(a) = F^+_{x,\psi}(a) - F^-_{x,\psi}(a).$$

Assume $t \in V^+$ and $x \in \mathcal{A}(D_t)$. Then the locality condition implies

$$F_{x,\psi}(a) = 0 \quad \text{for} \quad a \in (2D_t)'.$$

This implies that one can write

$$F_{x,\psi}(a) = G^+_{x,\psi}(a) - G^-_{x,\psi}(a)$$

with

$$\text{supp } G^+_{x,\psi}(a) \subset -2t + \overline{V}^+$$
$$\text{supp } G^-_{x,\psi}(a) \subset 2t - \overline{V}^+.$$

Let us denote the spectral projections of $U(a)$ by $E(\Delta)$. We say a vector $\psi \in \mathcal{H}$ has support in Δ if $E(\Delta)\psi = \psi$ holds. From the support property of $G^{\pm}$ it follows that $\mathcal{F}^{-1}G^+_{x,\psi}$ is a boundary value of an analytic function holomorphic in $T(V^-)$ and $\mathcal{F}^{-1}G^-_{x,\psi}$ is a boundary value of an analytic function holomorphic in $T(V^+)$. Hence one has to deal with an edge of the wedge problem in case that supp $\mathcal{F}^{-1}(G^+ - G^-)$ is not the whole of $\mathbb{R}^d$. For computing this support one uses the equation $F^+ - F^- = G^+ - G^-$. By the definition of these functions it follows that

$$\text{supp } \mathcal{F}^{-1}F^+_{x,\psi} \subset \text{spec } U$$
$$\text{supp } \mathcal{F}^{-1}F^-_{x,\psi} \subset 2\Delta - \text{spec } U,$$

and consequently

$$\text{supp } \mathcal{F}^{-1}(G^+_{x,\psi} - G^-_{x,\psi}) \subset \text{spec } U \cup \{2\Delta - \text{spec } U\}$$

where Δ is the support of ψ which is at the same time a subset of the spectrum of U. Notice that this support depends only on the set Δ and the spectrum of U and not on the special features on the operator x or the vector $\psi \in E(\Delta)\mathcal{H}$. Moreover, the fact that $\mathcal{F}^{-1}G^\pm$ are boundary values of analytic functions in $T(V^\mp)$ depends only on $x \in \mathcal{A}(D_t)$, but not on the choice of x and the choice of t, as long as t is a finite vector in V^+.

Let $G(\Delta)$ be the central support of $E(\Delta)$. Then the set of vectors $\{x\psi\}$ is total in $G(\Delta)\mathcal{H}$ provided ψ varies over $E(\Delta)\mathcal{H}\}$ and x over $\mathcal{A}(D_t)$ for arbitrary $t \in V^+$. This implies that investigating the functions $F^+_{x,\psi}$ with these conditions on x and ψ is equivalent to investigating $U(a)G(\Delta)$.

The edge of the wedge problem associated with the functions $\mathcal{F}^{-1}G^-$ and $\mathcal{F}^{-1}G^+$ has a coincidence domain

$$G = \mathbb{R}^d \setminus \{\text{spec } U \cup \{2\Delta - \text{spec } U\}\}$$

which in many cases can be enlarged to a domain $\tilde{G}$. We write this as

$$\tilde{G} = \mathbb{R}^d \setminus \Gamma \quad \text{with} \quad \Gamma \subset \text{spec } U \cup \{2\Delta - \text{spec } U\}.$$

Clearly the two functions $\mathcal{F}^{-1}F^+_{x,\psi}$ and $\mathcal{F}^{-1}F^-_{x,\psi}$ coincide outside of the set $2\Delta - \text{spec } U$. As a consequence one obtains the result

IV.1.1 Proposition:

We stay with the assumptions made so far in this section. If one knows that the spectrum of $G(\Delta)U(a)$ is contained in $\widetilde{\text{spec}} U$, and if by the edge of the wedge problem the domain $\mathbb{R}^d \setminus \widetilde{\text{spec}} U \cup \{2\Delta - \widetilde{\text{spec}} U\}$ can be enlarged to $\mathbb{R}^d \setminus \Gamma$, then one has

$$\text{spec } G(\Delta)U(a) \subset \left\{\{\Gamma \cap \widetilde{\text{spec}} U\} \cup \{\widetilde{\text{spec}} U \cap (2\Delta - \widetilde{\text{spec}} U)\}\right\}.$$

Proof: From the discussion of the edge of the wedge problem we know supp $\mathcal{F}^{-1}F_{x,\psi} \subset \Gamma$. Furthermore we know $\mathcal{F}^{-1}F = \mathcal{F}^{-1}F^+ - \mathcal{F}^{-1}F^-$.

Since in the intersection of the two supports some accidental cancellation might occur we can only draw conclusions for that part of the spectrum of $G(\Delta)U(a)$ which is outside of $2\Delta - \widetilde{spec}\,U$. The rest is a consequence of the previous discussion. □

At first sight this proposition seems to be not very useful because of two reasons:

1) The coincidence domain and therefore also the envelope of holomorphy depends on the choice of the set Δ.

2) Statements about the spectrum of the translations can only be obtained for the points outside of $2\Delta - V^+$.

This difficulties can be overcome. Let us assume for a moment that we are dealing with a factor representation. In this case every $E(\Delta)$ which is non–zero has central support $\mathbb{1}$. The strategy will be to vary Δ in order to obtain statements which are true for all subsets Δ and which are therefore statements which are independent of Δ. This strategy works also if a center is present. In this case one has to determine sufficiently many sets such that $E(\Delta)$ has central support $\mathbb{1}$.

IV.2 Locality and minimal translations

Let $\{\mathcal{A}(O), \mathcal{A}, \mathbb{R}^d, \alpha\}$ be a theory of local observables. Recall that we denote by V^+ the open forward light-cone in configuration space $\mathbb{R}^d$ and by $\overline{V}^+$ the dual cone in momentum space. Denote by $E(\overline{V}^+)$ the projection defined in II.6.1 (iii). In section II.6 we have constructed in $\mathcal{A}^{**}E(\overline{V}^+)$ a continuous unitary representation of the translations which implements the automorphisms α_a and the spectrum of which is contained in $\overline{V}^+$. In the one-dimensional case we constructed, in section II.4, a special representation which was called minimal. Now we want to generalize this concept to the d-dimensional situation. However, we start with some general considerations which are based on an arbitrary convex, proper cone.

IV.2.1 Definition:

Let $\{\mathcal{A}, \mathbb{R}^d, \alpha\}$ be a C^*-dynamical system and let $C \subset \mathbb{R}^d$ be an open, convex cone with proper closure and denote by C' its (closed) dual cone. Let $U(a) \in \mathcal{A}^{**}E(C')$ be a continuous unitary representation of the translations with

(i) Spectrum $U(a) \subset C'$,

(ii) $U(a)$ implements the automorphism α_a.

Then $U(a)$ is called minimal, if for any other continuous, unitary representation $V(a) \in \mathcal{A}^{**}E(C')$ of the translations which also fulfils (i) and (ii), one has spectrum $V(a)U^*(a) \subset C'$.

Remarks:

(i) If we write $U(a) = \exp\{i(a, P)\}$ and $V(a) = \exp\{i(a, Q)\}$, then $U(a)$ minimal is equivalent to the relation $(t, P) \leq (t, Q)$ for every $t \in \overline{C}$.

(ii) If a minimal representation exists then it is necessarily unique.

(iii) If the cone C' is a simplicial cone then a minimal representation always exists. This can be seen by repeating the arguments of II.5.

(iv) If the cone C is general then the existence of a minimal representation can be expected only if there is an intimate relation between the algebraic structure and the cone C.

It is our aim to show that this is the case in the theory of local observables. To this end we remark first if $U(a)$ and $V(a)$ are two representations of the translations which fulfil the requirements of Definition IV.2.1, except minimality, then $V(a)U^*(a) \in \mathcal{Z}(\mathcal{A}^{**}E(C'))$. Hence we define:

IV.2.2 Definition:

(a) *Let $U(a)$ be a representation of the translations with the above require-ments and write $U(a) = \exp\{i(a, P)\}$, where $P = \{P_0, ..., P_{d-1}\}$ are the generators. Then we introduce the multi-component objects $\mathcal{P}(U)$ as the set of $Q = \{Q_0, ..., Q_{d-1}\}$ with self–adjoint Q_i, $Q_i = Q_i^*$ and*

(i) *$f(Q_i) \in \mathcal{Z}(\mathcal{A}^{**}E(C'))$, for all bounded continuous function on $\mathbb{R}^d$,*

(ii) *$(t, P + Q) \geq 0, \quad \forall t \in \overline{C}$.*

(b) *In $\mathcal{P}(U)$ we introduce a t–semi–order by putting*

$$Q^1 \overset{t}{<} Q^2, \quad \text{if} \quad t \in \overline{C} \quad \text{is fixed and if} \quad (t, Q^1) \leq (t, Q^2),$$

where the last inequality is meant in the order of operators.

With this notation one obtains:

IV.2.3 Lemma:

The spaces $\mathcal{P} = \mathcal{P}(U)$ introduced above have the following properties:

(i) *Let $U^1(a) = \exp\{i(a, P^1)\}$ and $U^2(a) = \exp\{i(a, P^2)\}$ be two differ-ent representations fulfilling the spectrum condition and which belong to $\mathcal{A}^{**}E(C')$. Then $R = P^1 - P^2$ is affiliated with $\mathcal{Z}(\mathcal{A}^{**}E(C'))$ and one has:*

$$\mathcal{P}(U^1) = \mathcal{P}(U^2) + R.$$

(ii) *$\mathcal{P}$ is convex and if Q is affiliated with $\mathcal{Z}(\mathcal{A}^{**}E(C'))$ such that $(t, Q) \geq 0$ for every $t \in \overline{C}$ then $Q \in \mathcal{P}$.*

(iii) *For a given t–semi–order $\mathcal{P}$ is a lattice; namely if $Q^1, Q^2 \in \mathcal{P}$ then there exist*

$$Q_1 \in \mathcal{P}, \quad Q_1 \overset{t}{<} Q^1, \quad Q_1 \overset{t}{<} Q^2$$

and

$$Q_u \in \mathcal{P}, \quad Q^1 \overset{t}{<} Q_u, \quad Q^2 \overset{t}{<} Q_u.$$

(iv) *The bounded part of* $\mathcal{P}$ *(this means the set of all* $Q \in \mathcal{P}$ *such that* $\sum \|Q_i\| < \infty$ *) is closed in the ultra-weak operator topology of* $\mathcal{Z}(\mathcal{A}^{**}E(C'))$.

(v) *Let* $U(a) = \int_{C'} e^{i(a,p)} dE(p)$ *be the spectral representation of* $U(a)$. *Let* $\Delta \subset C'$ *be compact and let* $G(\Delta)$ *be the central support of* $E(\Delta)$. *Then* $\mathcal{P}G(\Delta)$ *is bounded below in every* t-*semi-order* $(t \in \overline{C'})$.

(vi) $U(a)$ *is minimal if and only if* $\mathcal{P}$ *contains only positive elements in every* t-*semi-order, namely,* $(t,Q) \geq 0$ *for every* $Q \in \mathcal{P}$ *and every* $t \in \overline{C}$.

Proof: (i) Assume $Q^1 \in \mathcal{P}(U^1)$. Then for $t \in \overline{C}$ one has $(t, P^2 + Q^1 + R) = (t; P^1 + Q^1) \geq 0$, hence $\mathcal{P}(U^1) + R \subset \mathcal{P}(U^2)$. Conversely if $Q^2 \in \mathcal{P}(U^2)$ then $(t, P^1 + Q^2 - R) = (t, P^2 + Q^2) \geq 0$ and therefore $\mathcal{P}(U^2) - R \subset \mathcal{P}(U^1)$.
(ii) Assume $Q^1, Q^2 \in \mathcal{P}$. Then for $t \in \overline{C}$ we have $(t, P + \lambda Q^1 + (1-\lambda)Q^2) = \lambda(t, P + Q^1) + (1-\lambda)(t, P + Q^2) \geq 0$ and hence $\lambda Q^1 + (1-\lambda)Q^2 \in \mathcal{P}$ for $0 \leq \lambda \leq 1$. If $(t,Q) \geq 0$ for every $t \in \overline{C}$ then clearly $(t, P+Q) \geq 0$ and hence $Q \in \mathcal{P}$.
(iii) Let $t \in \overline{C}$ be fixed and let Q^1 and Q^2 both be in $\mathcal{P}$. Then $(t, Q^1) - (t, Q^2)$ is self-adjoint and one can write $(t, Q^1) - (t, Q^2) = \Delta^+ - \Delta^-$ where $\Delta^{\pm} \in \mathcal{Z}(\mathcal{A}^{**}E(C'))$ are both positive. Let G^+ be the support projection of Δ^+ and put $Q_l = G^+ Q^2 + (E(C') - G^+)Q^2$ and $Q_u = G^+ Q^1 + (E(C') - G^+)Q^2$. It is clear that (t, Q_l) minimizes (t, Q^1) and (t, Q^2) and that (t, Q_u) majorizes the two operators. For arbitrary $t' \in \overline{C}$ one finds $(t', P + Q_l) = (t', G^+(P + Q^2)) + (t', (E(C') - G^+)(P + Q^1)) \geq 0$ by the choice of Q^1 and Q^2. The same argument holds for Q_u.
(iv) Let $Q^\alpha \in \mathcal{P}$ converge σ-weakly to Q. Then we have for every $\mathcal{A}^{**}E(C')$ normal state, $\lim_\alpha \omega((t, P + Q^\alpha)) = \omega((t; P + Q)) \geq 0$. Since this holds for every $t \in \overline{C}$ it implies $Q \in \mathcal{P}$.
(v) If $\Delta \subset C'$ is compact and $E(\Delta) \neq 0$ we denote by $\lambda_t = \sup\{(p, t); p \in \Delta\}$ which is bounded since Δ is compact. Let $p_0 \in \Delta$ such that $(p_0, t) = \lambda_t$; then $\Delta - p_0 \cap C' = \{0\}$ and consequently for every $Q \in \mathcal{P}$ we have $(t, (Q + p_0)G(\Delta)) \geq 0$ which implies $-p_0 G(\Delta) \overset{t}{<} QG(\Delta)$.
(vi) This is trivially equivalent to the Definition IV.2.1 of minimality. □

Next we want to show the existence of minimal elements in the t-order for any $t \in \overline{C}$.

IV.2.4 Proposition:

Let $\{\mathcal{A}, \mathbb{R}^d, \alpha, C'\}$ *be a* C^*-*dynamical system and* $U(a)$ *a continuous representation of the translation group. Let* $\mathcal{P}$ *be the set defined in* IV.2.2, *and let* $t \in \overline{C}$. *Then*

(1) $\mathcal{P}$ *contains minimal elements with respect to the order* $\overset{t}{<}$.

(2) *For every $Q \in \mathcal{P}$ and every t–minimal minimal element Q' one has $Q' \overset{t}{<} Q$.*

(3) *If Q is a minimal element with respect to the order $\overset{t}{<}$ then* spec $Q \subset \{p; (p,t) \leq 0\}$.

Proof: Let $\Delta \in C'$ be a bounded set such that $E(\Delta) \neq 0$, where $E(\Delta)$ denotes the spectral projection of $U(a)$. Let $G(\Delta)$ be the central support of $E(\Delta)$. Then, by Lemma IV.2.3(v), the spectrum of the family of elements $\{QG(\Delta); Q \leq 0\}$ is contained in a common bounded set, and hence $\sum \|Q_i\| \leq M < \infty$. Let Q_β be a monotone t-decreasing net in this set. Since the unit ball in $\mathcal{Z}G(\Delta)$ is weakly compact there exists a convergent subnet Q_γ with limit Q' which belongs to $\mathcal{P}G(\Delta)$ by Lemma IV.2.3 (iv). Since Q_β was decreasing one has $Q' \overset{t}{<} Q_\beta$. Using Zorn's Lemma we see that $\mathcal{P}G(\Delta)$ contains minimal elements in the t-semi-order. Moreover, for every $QG(\Delta)$ and every minimal element Q' one has $Q'G(\Delta) \overset{t}{<} QG(\Delta)$ because $\mathcal{P}G(\Delta)$ is a lattice in the t-semi-order. The sets $G_n = G(\{p \in C'; (p,t) \leq n\})$ have the property that G_n converges to $E(C')$. Define

$$M_t(\mathcal{P}) = \{\sum_1^\infty Q_n (G(n) - G(n-1)); \text{ with } Q_n \text{ a } t\text{–minimal element of}$$

$$\mathcal{P}(G(n) - G(n-1))\}.$$

If we have $Q \in \mathcal{P}$ and $Q' \in M_t(\mathcal{P})$ then by construction $Q'(G(n) - G(n-1)) \overset{t}{<} Q(G(n) - G(n-1))$ and hence $Q' \overset{t}{<} 0$. This shows (1) and (2). Finally let Q be t-minimal. Then $Q \overset{t}{<} 0$ which implies $(Q,t) \leq 0$. Hence the spectrum of Q is contained in the half-space $(p,t) \leq 0$. □

Up to now the condition of locality has not been used; however it becomes essential for the next result.

IV.2.5 Proposition:

Let $\{\mathcal{A}(O), \mathcal{A}, \mathbb{R}^d, \alpha\}$ be a theory of local observables and let $U(a) \in \mathcal{A}^{**}E(\overline{V}^+)$ be a continuous unitary representation of the translations fulfilling the spectrum condition and implementing α_a. Let $\mathcal{P}$ be as defined in IV.2.2. Let $Q \in \mathcal{P}$ and assume that the spectrum of Q contains points which are spacelike. Then for every given $t \in V^+$ there exists $Q_t \in \mathcal{P}$ with $Q_t \overset{t}{<} Q$ but $Q_t \neq 0$ and $Q_t \neq Q$.

Proof: Let $p \in$ spec Q with $p^2 < 0$. Write $p = b + \lambda t$ with $(b,t) = 0$ and $\lambda^2 < \frac{-b^2}{t^2}$. This is possible since t was timelike. Choose ϵ small enough such that $(|\lambda| + \epsilon)^2 < \frac{-b^2}{t^2}$. Then the double cone $D_{p-\epsilon t, p+\epsilon t}$ is still contained in the spacelike points and $G(D) := G(D_{p-\epsilon t, p+\epsilon t}) \neq 0$, where G is the spectral projection of Q. From the definition of Q it follows that the spectrum of $(P+Q)G(D)$ is contained in $\overline{V}^+ \cap \{(p-\epsilon t)+\overline{V}^+\}$. See Fig. 3. Now let $E(\lambda)$ be the spectral family of $(P+Q,t)$. Choose λ_0 such that $E(\lambda_0) \neq 0$ and let $G(\lambda_0)$ be the central support of $E(\lambda_0)$. Denote by Δ_n the set $\Delta_n = \{q \in \overline{V}^+ \cap \{(p-\epsilon t)+\overline{V}^+\}; (q,t) \leq n\lambda_0\}$, $n = 1,2,....$ Since $(p-\epsilon t)$ is spacelike one sees that there exists $\sigma < 0$ such that $\Delta_2 + \sigma b$ is contained in the interior of V^+ and hence one can find $\tau > 0$ such that $\Delta_2 + \sigma b - \tau t \subset \overline{V}^+$.

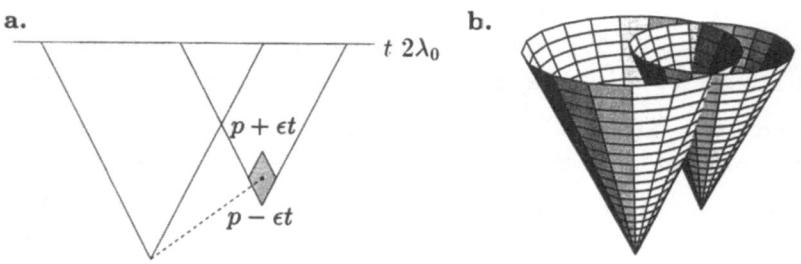

Fig. 3. Example for the construction of the minimal translations.
a. The support of $P+Q$ if Q contains a spacelike point.
b. The three–dimensional picture of the situation.

Now we use the technique described in detail in section IV.1 and apply the Jost-Lehmann-Dyson representation (Section III.4) to the coincidence domain

$$\mathbb{R}^d \setminus \left((\overline{V}^+ \cap ((p-\epsilon t)+\overline{V}^+) \cup \{2\Delta_1 - (\overline{V}^+ \cap ((p-\epsilon t)+\overline{V}^+)\}.$$

This gives an enlargement of the coincidence domain to $\mathbb{R}^d \setminus K + \{\overline{V}^+ \cup -\overline{V}^+\}$ where K is a closed subset of Δ_2, namely $V^+ \cap ((p-\epsilon t + \overline{V}^+) \cap \{2\Delta_1 - (\overline{V}^+ \cap ((p-\epsilon t)+\overline{V}^+))\}$. But this implies by Proposition IV.1.1 that $(P+Q)G(D)G(\lambda_0)$ has its spectrum in $\Delta_2 + \overline{V}^+$. Consequently spec $(P+Q)G(D)G(\lambda_0) + \sigma b - \tau t$ is contained in $\overline{V}^+$. Define

$$Q_t = Q(1 - G(D)G(\lambda_0)) + (Q + \sigma b - \tau t)G(D)G(\lambda_0).$$

By construction one has $Q_t \in \mathcal{P}$ and since $\tau > 0$ one has $Q_t \neq Q$ and $Q_t \overset{t}{<} Q$. $\qquad\square$

Combine this result with Proposition IV.2.4 to obtain

IV.2.6 Theorem:

Let $\{\mathcal{A}(O), \mathcal{A}, \mathbb{R}^d, \alpha\}$ be a theory of local observables, and let $U(a)$ be a continuous representation of the translation group satisfying the spectrum condition and implementing α_a. Let $\mathcal{P}$ be the set defined in IV.2.2. Then $\mathcal{P}$ contains a unique element which is minimal with respect to every t-semi-order, $t \in V^+$.

Proof: Take $t \in V^+$ and choose a t-minimal element $Q \in \mathcal{P}$. Since $0 \in \mathcal{P}$ it follows from IV.2.4 (2) that $Q \overset{t}{<} 0$. This implies that the spectrum of Q contains only points in $\overline{V}^-$ or spacelike points. If spec Q contains spacelike points then by IV.2.5 it is not minimal. Hence spec $Q \subset \overline{V}^-$.

Now let $t_1, t_2 \in V^+$ be two linearly independent vectors and let Q_1 and Q_2 be t_1 and t_2 minimal respectively. Define $P_i = P - Q_i$, $i = 1, 2$. Then according to Lemma IV.2.3 (i) one has $\mathcal{P}(P_i) = \mathcal{P}(P) - Q_i$. But this implies $Q_1 - Q_2$ is t_1 minimal in $\mathcal{P}(P_2)$ and $Q_2 - Q_1$ is t_2 minimal in (P_1). Consequently we have the inclusions spec $(Q_1 - Q_2) \subset -\overline{V}^+$ and spec $(Q_2 - Q_1) \subset -\overline{V}^+$ and hence spec $(Q_1 - Q_2) = \{0\}$ or $Q_1 = Q_2$. Since t_2 was arbitrary one sees that Q_1 is minimal with respect to every t-order, $t \in V^+$. From this one also concludes its uniqueness. □

Notice that this minimal element is also minimal with respect to all $t \in \overline{V}^+$. This can easily be seen by standard limiting procedures.

IV.2.7 Corollary:

Let $\{\mathcal{A}(O), \mathcal{A}, \mathbb{R}^d, \alpha\}$ be a theory of local observables. Then there exists a unique continuous unitary group representation $U(a) \in \mathcal{A}^{**} E(\overline{V}^+)$ which fulfils the spectrum condition and which is minimal in the sense of Definition IV.2.1.(ii).

Proof: Choose any continuous unitary group representation $V(a) \in E(\overline{V}^+)\mathcal{A}^{**}$. Let Q be the unique minimal element in $\mathcal{P}(V)$ and define $U(a) = V(a) \exp\{i(Q, a)\}$. By Lemma IV.2.3 (i) one has $\mathcal{P}(U) = \mathcal{P}(V) - Q$. Since Q was minimal in $\mathcal{P}(V)$ one has $Q \overset{t}{<} Q'$ for every Q' in $\mathcal{P}(U)$. This shows that U is minimal. Let U' also be minimal. Then one has spec $U(a)U'^*(a) \subset \overline{V}^+$ and spec $U'(a)U^*(a) \subset \overline{V}^+$ which gives $U'(a) = U(a)$. □

IV.3   Locality and the shape of the spectrum

Let $\{\mathcal{A}(O), \mathcal{A}, \mathbb{R}^d, \alpha\}$ be a local ring system and let $\{\pi, \mathcal{H}, U(a)\}$ be a faithful normal representation of $\mathcal{A}^{**}E(\overline{V}^+)$. In the following we will identify $U(a)$ with the unique minimal representation in $\mathcal{A}^{**}E(\overline{V}^+)$ which exists according to Corollary IV.2.7. This section is devoted to the investigation of the spectrum of $U(a)$.

Since $U(a)$ is uniquely defined one can give its generators an absolute meaning and call them the energy and momentum operators of the theory. This new feature will be exploited in this section. In particular it will turn out that the interplay between the spectrum condition and the locality assumption is closely related to a particle interpretation of the theory.

We start the investigations with the following observations:

IV.3.1 Lemma:

Take the assumptions described in the beginning of this section. Denote the spectral projections of $U(a)$ by $E(\Delta)$. Let $b \in \mathbb{R}^d$ with $b^2 < 0$ and let

$$\Delta = \overline{V}^+ \setminus (b + \overline{V}^+).$$

Then $E(\Delta)$ has central support $\mathbb{1}$ (in $\mathcal{A}^{**}E(\overline{V}^+)$).

Proof: Let G be the central support of $E(\Delta)$. Then $U(a)(\mathbb{1} - G)$ has its spectral support in $\overline{V}^+ \cap (b + \overline{V}^+)$ and consequently $-(\mathbb{1} - G)b \in \mathcal{P}(U)$. Since U is minimal one has spec $Q \subset \overline{V}^+$ for every $Q \in \mathcal{P}(U)$ and hence $(\mathbb{1} - G) = 0$ since b is spacelike.   □

With help of this result one proves

IV.3.2 Lemma:

In addition to the assumptions described in the beginning of this section consider a projection $G \in \mathcal{Z}(\mathcal{A}^{**}E(\overline{V}^+))$ and let $p \in \partial V^+$, $p \neq 0$. Assume that p does not belong to the spectrum of $U(a)G$:
(a) If $d > 2$ then no point $p \in \partial V^+$ with $p \neq 0$ belongs to spec $U(a)$.
(b) If $d = 2$ then no point λp, $\lambda > 0$ belongs to spec $U(a)G$.

Proof: Since spec $U(a)G$ is closed it follows that with p a whole neighbourhood $\mathcal{U}$ of p lies outside the spectrum. Choose a timelike direction p_0 and a spacelike direction p_1 with $(p_0, p_1) = 0$ and $p_0^2 + p_1^2 = 0$. For every $\epsilon > 0$ Lemma IV.3.1 implies that (spec $U(a)G) \cap \complement (\epsilon p_1 + \overline{V}^+) = \Delta$ is not empty. We now apply to this situation the technique described in detail in section VI.1. To obtain an enlargement of the domain of holomorphy and hence a statement about the spectrum we use the Jost-Lehmann-Dyson method described in Theorems III.4.5 and III.4.7. This implies that the part of the spectrum of $U(a)G$ which lies in $(2\epsilon p_1 + \overline{V}^+)$ belongs to the union of those hyperboloids with center in $\overline{V}^+ \cap \complement(2\epsilon p_1 + \overline{V}^+)$ which do not enter the neighbourhood $\mathcal{U}$ of p. This shows in particular that a neighbourhood of $\{\lambda p; \lambda > 2\epsilon\}$ does not belong to spec $U(a)G$. See Fig.4. From this one obtains the results for $d = 2$ since the above statement holds for every $\epsilon > 0$. If the dimension of the space is larger than two then one has to vary p_1 and ϵ in such a way that $q \in \partial V^+, q \neq 0$ does not belong to any hyperboloid used by the J.L.D. representation. This is possible since the hyperboloids are not allowed to enter the neigbourhood $\mathcal{U}$ of the point p. We learn by the compactness of the set of directions that no point of ∂V^+ belongs to spec $U(a)G$ with the possible exception of point zero. $\qquad\square$

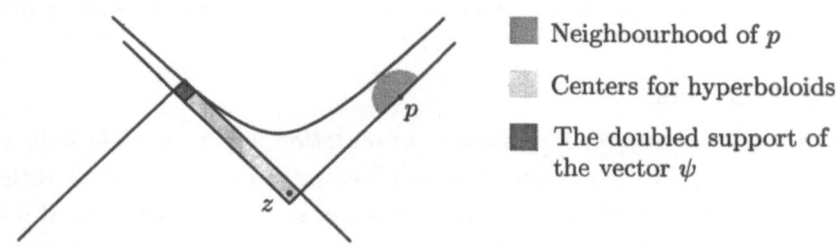

Fig. 4. Sample hyperboloid with center z if the point $p \in \partial V^+$ does not belong to the spectrum of the translations.

As a consequence of this result we obtain in addition

IV.3.3 Corollary:

Keep the assumptions and notations of the last lemma. If $\partial V^+ \setminus \{0\}$ does not belong to spec $U(a)G$ and if 0 belongs to spec $U(a)G$, then $\{0\}$ is an isolated point of the spectrum.

Proof: Let $p_0 \in V^+$ with $p_0^2 = 1$. Then for every $p_1 \in p_0^\perp$ with $p_1^2 = -1$ one has $p_0 + p_1 \notin$ spec $U(a)G$. Since the ball $p_1^2 = -1$ is compact there exists $\delta > 0$ such that $p_0 + (1 - \delta)p_1 \notin$ spec $U(a)G$. By Lemma IV.3.1 again spec $U(a)G \cap \complement(\epsilon p_1 + \overline{V}^+) = \Delta$ is not empty and we apply to this the technique described in section IV.1. However, to obtain an enlargement of the domain of holomorphy in this case we use the double cone Theorem III.3.2 with respect to the cone V^+ and obtain that the union of the interior of the double cones $\langle 2\epsilon(p_0 + p_1), p_0 + (1 - \delta)p_1 \rangle$ does not belong to the spectrum. Taking the limit $\epsilon \to 0$ we see that zero is an isolated point of the spectrum.$\square$

Another consequence we mention for later use is

IV.3.4 Corollary:

Keep the assumptions of Lemma IV.3.2. Assume $d = 2$. Let p_0, p_1 be fixed such that $p_0 \in V^+$, $p_0^2 = 1$, $p_1^2 = -1$ and $(p_0, p_1) = 0$. Assume $\{\lambda(p_0 + p_1), \lambda > 0\}$ does not belong to spec $U(a)G$, but that $\{\lambda(p_0 - p_1), \lambda \geq 0\}$ belongs to spec $U(a)G$. Then there exists an open neighbourhood $\mathcal{U}$ of zero such that

$$\text{spec } U(a)G \cap \mathcal{U} = \{\lambda(p_0 - p_1), \ \lambda \geq 0\} \cap \mathcal{U}.$$

Proof: The proof of this result uses exactly the same method as used in the proof of Corollary IV.3.3. Therefore we don't need to repeat it. □

In order to formulate the first result about the shape of the spectrum of translations we have to define what is meant by the lower boundary of the spectrum.

IV.3.5 Definition:

Let $U(a)$ be a continuous unitary representation of the translations such that spec $U(a) \subset \overline{V}^+$. A point p in the boundary of spec $U(a)$ is called a lower boundary point if there exists a continuous curve C starting at p and ending at some fixed point q lying in the interior of $-V^+$ such that $C \cap$ spec $U(a) = \{p\}$.

With this definition one obtains

IV.3.6 Proposition:

Let $\{\mathcal{A}(O), \mathcal{A}, \mathbb{R}^d, \alpha\}$ be a theory of local observables and assume $U(a)$ is the unique minimal representation of the translations described in section IV.2. Then the lower boundary of the spectrum of $U(a)G$ is a Lorentz invariant set for every projection $G \in \mathcal{Z}(\mathcal{A}^{**}E(\overline{V}^+))$.

Proof: There are several different cases which need different considerations:

Case 1: ∂V^+ belongs to spec $U(a)G$. In this case ∂V^+ is the lower boundary which is obviously invariant.

Case 2: No point of ∂V^+ belongs to spec $U(a)G$.

Case 3: $0 \in$ spec $U(a)G$, but no other point of ∂V^+ belongs to spec $U(a)G$.

Case 4: This can happen only for $d = 2$. Let $p_0 \in V^+$ and $(p_1, p_0) = 0$ with $p_0^2 = 1$, $p_1^2 = -1$. Assume that $\{\lambda(p_0 + p_1); \lambda > 0\}$ does not belong to spec $U(a)G$ but $\{\lambda(p_0 - p_1); \lambda \geq 0\} \subset$ spec $U(a)G$. See Fig.5 next page.

From Lemma IV.3.2 it follows that these cases describe all possibilities. First we investigate the second case. Since zero does not belong to

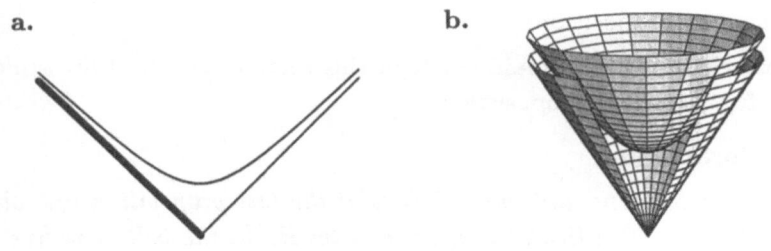

a.

b.

Fig. 5. Sample boundaries for the minimal spectrum.
a. In two dimensions: The massive hyperboloid and the spectrum of the "left movers".
b. The massless and the massive hyperboloids in three dimensions.

spec $U(a)G$ there exists a whole neighbourhood of zero which does not belong to the spectrum. Choose a vector $p_0 \in V^+$ with $p_0^2 = 1$. Then there is a constant $m > 0$ such that $(mp_0 + p_0^\perp)$ is a supporting hyperplane of spec $U(a)G$. Let $p_1 \in p_0^\perp$ with $p_1^2 = -1$ and $\Delta = \text{spec } U(a)G \cap \complement(\epsilon p_1 + V^+)$. Then we apply the technique described in IV.1 with vectors $\psi \in E(\Delta)\mathcal{H}$ where we use the Jost-Lehmann-Dyson formula. One sees that in the limit $\epsilon \to 0$ one obtains a set Γ_{p_1} with spec $U(a)G \subset \Gamma_{p_1}$ where Γ_{p_1} is the intersection of $\overline{V}^+$ with the union of all hyperboloids $h(l, m)$ with apex at points $l \in \partial V^+$ with $l^2 = 0$ and $(l, p_1) \geq 0$, such that their upper branch is above the supporting hyperplane $mp_0 + p_0^\perp$.

Now changing the direction of p_1 one sees by taking intersections that spec $U(a)G \subset \{p; \ p^2 \geq m^2\} \cup 0$. Since we know that zero is not a point of the spectrum we have spec $U(a)G \subset \{p; \ p^2 \geq m^2\}$. On the other hand, since $mp_0 + p_0^\perp$ was a supporting hyperplane the lower boundary of spec $U(a)G$ has to be $\{p; \ p^2 = m^2\}$. Otherwise one could change the direction of p_0 and obtain a contradiction to the value of m. In the third case one can argue as before. The only difference is the point zero. It cannot be removed by taking the intersection. Therefore zero remains a point of the spectrum. Finally the third case can be treated as follows. We know that $\{p_0^2 = p_1^2\} \cap \{p_1 \leq 0\}$ belongs to the spectrum by assumption and Lemma IV.3.2. On the other hand we can apply the method described in section IV.1 to the function $\Box F_{\psi,x}^+(a)$ and $\Box F_{\psi,x}^-(a)$. By Corollary IV.3.4 we know that there exists a neighbourhood of zero not belonging to spectrum of $\Box U(a)$ and hence by Corollary IV.3.3 we find that no point of $\{p \ p^2 = 0\}$ belongs to the spectrum of $\Box U(a)$. Applying now the result of case 2 we find that there exists an

$m > 0$ such that the lower boundary consists of two parts, the hyperboloid $\{p; \; p^2 = m^2, m > 0\}$ and one half of the zero hyperboloid $p^2 = 0$. This establishes the proposition. $\qquad\qquad\square$

Before coming to the main result of this section we will draw some conclusions from the last proposition:

IV.3.7 Corollary:

With the assumptions and the notation of the last proposition one obtains: Let $l \in \partial V^+$, $l \neq 0$ a lightlike vector and let B_r be the ball of radius r and center 0. Denote by $\Delta(l,r)$ the set $\left(\bigcup_{\lambda \geq 0} \{\lambda l + B_r\} \right) \cap \overline{V}^+$. Then for every projection $G \in \mathcal{Z}(\mathcal{A}^{**} E(\overline{V}^+))$ and for every $\epsilon > 0$ one has: Central support of $E(\Delta(l, \epsilon))G = G$, where $E(\Delta)$ denotes the spectral projection of the unique minimal representation in $\mathcal{A}^{**} E(\overline{V}^+)$.

Proof: Let F be the central support of $E(\Delta(l, \epsilon))G$ and assume $F \neq G$. Then $(G - F)E(\Delta(l, \epsilon))G = 0$. Since the lower boundary of the spectrum of $(G - F)U(a)$ is invariant there exists a hyperboloid $M > 0$ such that $\{p; \; p^2 = M^2\}$ is the lower boundary (or is part of it). Choose a co-ordinate system such that $l = (1, 1)$ with $(1)^2 = 1$. Let $l' = (1, -1)$. Then $\lambda l + \alpha l' \in \Delta(l, \epsilon)$ for $\lambda > 0$ and $0 \leq \alpha \leq \frac{\epsilon}{\sqrt{2}}$. We have $(\lambda l + \alpha l')^2 = 4\lambda\alpha$. Choosing $\alpha = \frac{\epsilon}{2}$ and $\lambda \geq \frac{M^2}{2\epsilon}$ we see that $(\lambda l + \alpha l')^2 \geq M^2$. This shows that $\Delta(l, \epsilon) \cap$ spec $(G - F)U(a) \neq 0$ which is a contradiction. $\qquad\square$

Now we are prepared for the main result of this section:

IV.3.8 Theorem:

Let $\{\mathcal{A}(O), \mathcal{A}, \mathbb{R}^d, \alpha\}$ be a theory of local observables and assume that $U(a)$ is the unique minimal representation of the translations in $\mathcal{A}^{**} E(\overline{V}^+)$ described in section IV.2. Then for every projection $G \in \mathcal{Z}(\mathcal{A}^{**} E(\overline{V}^+))$, spec $U(a)G$ is a set which is invariant under Lorentz transformation.

Proof: From Proposition IV.3.6 we know that the lower boundary of the spectrum of $U(a)G$ is a set which is invariant under Lorentz transformations. Hence it remains to show that holes are also invariant under Lorentz transformations. To this end we have to combine the technique of section IV.1 with the results developed in the sections 5 and 6 of chapter III. To this end our coincidence domain will consist of two parts. G_1 will be the points which are space-like with respect to an order interval: $G_1 = < 0, 2p >'$, $p \in V^+$. The domain G_2 will be a double cone. We start with the assumption that there is a hole in the spectrum of $U(a)G$. Then by combining the method of section IV.1 with the double cone Theorem III.3.2 with respect to the cone V^+ one can assume that the hole is the union of order intervals. It is clear that the theorem holds true if we can prove it under the assumption that the starting hole is an order interval $< b, c >^\circ$ with $b^2 = m_1^2$, $c^2 = m_2^2$, and $m_1 < m_2$.

Fix a coordinate system and let $l = (1, 1)$ with $\mathbf{l}^2 = 1$ be a lightlike vector and let $l' = (1, -1)$. Then there exists $\epsilon > 0$ such that $(2p - \overline{V}^+) \cap$ $< b, c >= \emptyset$ for $p = \lambda l + \epsilon l'$. From Corollary IV.3.7 we know that for every $G \in \mathcal{Z}(\mathcal{A}^{**} E(\overline{V}^+))$ there exists $\lambda > 0$ such that $E(< 0, p >)G \neq 0$ and that the central projection of $E(< 0, p >)$ tends to G for $\lambda \to \infty$. Hence the method of section IV.1 and Theorem III.6.3 implies that the spectrum of $U(a)G$ vanishes in $\widetilde{G}_2$, where $\widetilde{G}_2$ is defined in Definition III.6.2. The set $\widetilde{G}_2$ depends on the vector $p = \lambda l + \epsilon l'$. Let $\widehat{G}_2(l) = \cup_\epsilon \{p; p = \lambda l + \epsilon l'\}$. For λ sufficiently large $\widetilde{G}_2$ becomes independent of λ so that one can compute the boundary of $\widehat{G}_2(l)$. This is given by four hyperboloids:

(i) $h(0, \sqrt{b^2})$,
(ii) $h(0, \sqrt{c^2})$,
(iii) $h(\lambda_b l, 0)$ where $\lambda_b l = \{\lambda l\} \cap \{b - \partial V^+\}$,
(iv) $h(\lambda_c l, 0)$ where $\lambda_c l = \{\lambda l\} \cap \{c - \partial V^+\}$.
See Fig. 6.

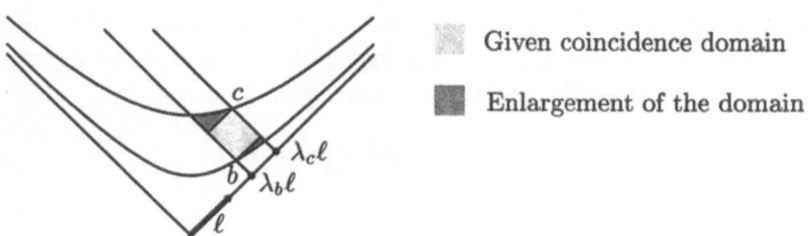

        Given coincidence domain

        Enlargement of the domain

Fig. 6. First step of the enlargement of the hole in the spectrum with help of the analyticity obtained from the spacelike points.

From this we see that in $\widehat{G}_2^\circ$ there are new order intervals $< b', c' >^\circ$ with $b'^2 = b^2$ and $c'^2 = c^2$. Changing the direction of l and repeating the argument with the new order intervals $< b', c' >$ one sees that the only set stable under this construction is

$$S(\sqrt{b^2}, \sqrt{c^2}) = \{p; \ b^2 < p^2 < c^2, \quad p_0 > 0\}.$$

This proves the Theorem. $\qquad\qquad\qquad\qquad\qquad\qquad\qquad\qquad\qquad\qquad\qquad\quad \square$

IV.4 Invariant states and the cluster property

In this section vacuum states are defined and some of their properties which are consequences of the locality condition are derived. Many of the results do not use the spectrum condition. First some preparation:

IV.4.1 Lemma:

Let $\{\mathcal{A}(O), \mathcal{A}, \mathbb{R}^d, \alpha\}$ be a local net (fulfilling the axioms I, II, and III) and let $a \in \mathbb{R}^d$ be spacelike. Then for every $x, y \in \mathcal{A}$ one has

$$\lim_{\lambda \to \infty} \| [x, \alpha_{\lambda a}(y)] \| = 0.$$

Proof: Since $\cup \mathcal{A}(O)$ is norm-dense in $\mathcal{A}$ one can find, for any given $\epsilon > 0$, bounded open sets O_1, O_2 and $u \in \mathcal{A}(O_1)$, $v \in \mathcal{A}(O_2)$ with $\|u - x\| < \frac{\epsilon}{4\|y\|}$ and $\|v - y\| < \frac{\epsilon}{4\|x\|}$. Since a is spacelike there exists λ_0 such that O_1 and $O_2 + \lambda a$ are spacelike for $\lambda > \lambda_0$. For such λ one has $[u, \alpha_{\lambda a}(y)] = 0$, and hence $[x, \alpha_{\lambda a}(y)] = [x, \alpha_{\lambda a}(y-v)] + [(x-u), \alpha_{\lambda a}(v)]$ implies $\| [x, \alpha_{\lambda a}(y)] \| < \epsilon$. Since ϵ is arbitrary one obtains the result. $\qquad\square$

This last result can be extended in the following way:

IV.4.2 Lemma:

Let $\{\mathcal{A}(O), \mathcal{A}, \mathbb{R}^d, \alpha\}$ be a local net, and let π be a representation of $\mathcal{A}$ on $\mathcal{H}$. Denote by $\mathcal{Z}_\pi$ the center of π'' and by $\mathcal{Z}'_\pi$ the commutant of $\mathcal{Z}_\pi$. Assume $\psi, \varphi \in \mathcal{H}$, $x \in \mathcal{A}$, $A \in \mathcal{Z}'_\pi$, and let a be spacelike. Then one has

$$\lim_{\lambda \to \infty} \big(\psi, [\pi(\alpha_{\lambda a} x), A] \varphi \big) = 0.$$

Proof: The set of operators of the form $\sum_i \pi(y_i) B_i$, $y_i \in \mathcal{A}$ and $B_i \in \pi(\mathcal{A})'$ is strongly dense in $\mathcal{Z}'_\pi$. Denote this algebra of finite sums by $\mathcal{B}$. If $A \in \mathcal{Z}'_\pi$ is selfadjoint then there exists a selfadjoint $B \in \mathcal{B}$ such that $\|(A - B)\varphi\| < \frac{\epsilon}{2\|x\|}$ and $\|(A - B)\psi\| < \frac{\epsilon}{2\|x\|}$. Hence one obtains

$$\lim_{\lambda \to \infty} |(\psi, [\pi(\alpha_{\lambda a} x), A]\varphi)| < \epsilon \|\psi\| \, \|\varphi\| +$$
$$\lim_{\lambda \to \infty} |(\psi, [\pi(\alpha_{\lambda a} x), B]\varphi)| = \epsilon \|\psi\| \, \|\varphi\|$$

since by Lemma IV.1.1 $\lim_{\lambda \to \infty} \| [\pi(\alpha_{\lambda a} x), \sum_i \pi(y_i) B_i] \| = 0$. As ϵ was arbitrary the result follows since every operator is a linear combination of selfadjoint ones. $\qquad\square$

Next the concept of an invariant state is needed.

IV.4.3 Definition:

Let $\{A, G, \alpha\}$ be a C^*-dynamical system. Then a state ω on A is called an invariant state if $\omega \circ \alpha_g = \omega$ for every $g \in G$.

IV.4.4 Lemma:

Let $\{A, G, \alpha\}$ be a C^*-dynamical system and let ω be an invariant state and $\{\pi, \mathcal{H}, \psi_\omega\}$ the G.N.S. representation of A defined by ω. Then there exists on $\mathcal{H}$ a unitary representation $U(g)$ of the group G with:

$$U(g)\psi_\omega = \psi_\omega \qquad x \in G,$$
$$U(g)\pi(x)U^*(g) = \pi(\alpha_g x), \quad x \in A.$$

Proof: Define $U(g)$ by the equation

$$U(g)\pi(x)\psi_\omega = \pi(\alpha_g x)\psi_\omega.$$

Since the representation is cyclic with cyclic vector ψ_ω one sees that $U(g)$ is densely defined and has a dense range. Because of the linearity of α_g, $U(g)$ itself is linear. Moreover, $\|U(g)\pi(x)\psi_\omega\|^2 = (\pi(\alpha_g x)\psi_\omega, \pi(\alpha_g x)\psi_\omega) = \omega(\alpha_g(x^*x)) = \omega(x^*x) = \|\pi(x)\psi_\omega\|^2$, since ω is an invariant state. Hence $U(g)$ is isometric and the first remark implies that $U(g)$ is unitary. From $\alpha_{g_1}\alpha_{g_2} = \alpha_{g_1 g_2}$ one calculates $U(g_1)U(g_2)\pi(x)\psi_\omega = U(g_1 g_2)\pi(x)\psi_\omega$. But this implies that $U(g)$ is a representation of the group G. $\qquad\square$

Next we want to combine Lemma IV.4.2 and Lemma IV.4.4.

IV.4.5 Proposition:

Let $\{A(O), A, \mathbb{R}^d, \alpha\}$ be a local net, ω be a translation invariant state, and $\{\pi, \mathcal{H}, \psi_\omega, U(a)\}$ the G.N.S. representation defined by ω with the canonical group representation given in the last lemma. Denote by E_0 the projection onto all vectors in $\mathcal{H}$ which are pointwise invariant under every $U(a)$. Assume in addition that $\mathcal{Z}_\pi$ is pointwise invariant under α_a, $a \in \mathbb{R}^d$, and let b be spacelike. Then the following relation holds for $x_i, y_j \in A$:

$$\lim_{\lambda \to \infty} (\psi_\omega, \pi(x_1)\pi(\alpha_{\lambda b}y_1)\pi(x_2)\cdots\pi(x_n)\pi(\alpha_{\lambda b}y_n)\psi_\omega)$$
$$= (\psi_\omega, \pi(x_1 x_2 \cdots x_n)E_0\pi(y_1 y_2 \cdots y_n)\psi_\omega)$$
$$= (\psi_\omega, \pi(y_i \cdots y_n)E_0\pi(x_1 \cdots x_n)\psi_\omega).$$

Proof: Repeated use of Lemma IV.4.1 gives

$$\lim_{\lambda \to \infty} \|\pi(x_1)\pi(\alpha_{\lambda a}y_1)\cdots\pi(x_n)\pi(\alpha_{\lambda a}y_n) - \pi(x_1...x_n)\pi(\alpha_{\lambda a}(y_1...y_n))\| = 0.$$

Since every $U(a)$ commutes with $\mathcal{Z}_\pi$ it follows that E_0 also belongs to $\mathcal{Z}_\pi'$. From the relation $E_0\psi_\omega = \psi_\omega$ and Lemma IV.4.2 one obtains

$$\lim_{\lambda \to \infty} (\psi_\omega, \pi(x_1...x_n)\pi(\alpha_{\lambda a}(y_1...y_n))\psi_\omega)$$
$$= \lim_{\lambda \to \infty} (\psi_\omega, \pi(x_1...x_n)E_0\pi(\alpha_{\lambda a}(y_1...y_n))\psi_\omega)$$
$$= \lim_{\lambda \to \infty} (\psi_\omega, \pi(x_1...x_n)E_0 U(\lambda a)\pi(y_1...y_n)\psi_\omega)$$
$$= (\psi_\omega, \pi(x_1...x_n)E_0\pi(y_1...y_n)\psi_\omega).$$

Since the commutation to the other side can be proved in the same manner, one obtains the desired result. □

In the last proposition we have seen that one can interchange the order of operators in the limit. This phenomenon will be investigated more closely in the last step of the preliminaries.

IV.4.6. Proposition:

Let $\{\mathcal{A}(O), \mathcal{A}, \mathbb{R}^d, \alpha\}$ be a local net and ω be an invariant state giving rise to the G.N.S. representation $\{\pi, \mathcal{H}, \psi_\omega, U(a)\}$. Denote by $\mathcal{R}$ the von Neumann algebra generated by $\pi(\mathcal{A})$ and the group representation $\{U(a); a \in \mathbb{R}^d\}$. Assume that the center $\mathcal{Z}_\pi$ commutes with $U(a)$. Then

(1) *E_0 is an abelian projection of $\mathcal{R}$ with central support $\mathbb{1}$ and consequently $\mathcal{R}$ is a von Neumann algebra of type I.*
(2) *The commutant of $\mathcal{R}$ is abelian and coincides with $\mathcal{Z}_\pi$.*

Remark:

This proposition remains true if the assumption that $\mathcal{Z}_\pi$ is pointwise invariant is dropped except for the modification that the commutant of $\mathcal{R}$ consist of the invariant elements of $\mathcal{Z}_\pi$. Since the more general result is not needed in what follows we refrain from proving it.

Proof: In the proof of Proposition IV.4.5 ψ_ω can be replaced by any other vector in $E_0\mathcal{H}$ which implies that one can iterate this proposition. Hence for $x_1, \cdots, x_4 \in \mathcal{A}$ one obtains:

$$(\psi_\omega, \pi(x_1)E_0\pi(x_2)E_0\pi(x_3)E_0\pi(x_4)\psi_\omega) =$$
$$(\psi_\omega, \pi(x_1)E_0\pi(x_3)E_0\pi(x_2)E_0\pi(x_4)\psi_\omega).$$

Since $\pi(\mathcal{A})\psi_\omega$ is dense in $\mathcal{H}$ it follows that $E_0\pi(\mathcal{A})\psi_\omega$ is dense in $E_0\mathcal{H}$. But this implies

$$[E_0\pi(x)E_0, E_0\pi(y)E_0] = 0 \quad \text{for} \quad x, y \in \mathcal{A}.$$

Next the set of operators $\sum_i \pi(x_i)U(a_i)$ is a $*$–algebra $\mathcal{R}_0$ since $\{\pi(x)U(a)\}^* = \pi(\alpha_{-a}x^*)U(-a)$ and $\pi(x)U(a)\pi(y)U(b) = \pi(x\alpha_a y)U(a+b)$. This is the algebra generated by $\pi(\mathcal{A})$ and $\{U(a)\}$ and it is dense in $\mathcal{R}$.

If $B \in \mathcal{R}_0$ with $B = \sum_i \pi(x_i)U(a_i)$ then one has $BE_0 = \sum_i \pi(x_i)E_0$ and hence $\mathcal{R}_0 E_0 = \pi(\mathcal{A})E_0$. From this one obtains $[E_0 B_1 E_0, E_0 B_2 E_0] = 0$ for $B_1, B_2 \in \mathcal{R}_0$. Notice next that $\mathcal{R}_0$ is strongly dense in $\mathcal{R}$ and that one obtains a weak limit point when going to a strong limit point with the two operators B_1 and B_2. Hence $E_0 \mathcal{R} E_0$ is an abelian algebra (acting on $E_0 \mathcal{H}$). Since now E_0 is the projector onto the common eigenstate of all $U(a)$ to the eigenvalue 1, it follows that E_0 belongs to the von Neumann algebra generated by $\{U(a)\}$ and consequently all the more so to $\mathcal{R}$. But this shows that E_0 is an abelian projection of $\mathcal{R}$. Furthermore from $E_0 \psi_\omega = \psi_\omega$ and $\overline{\mathcal{R}\psi_\omega} \supset \overline{\pi(\mathcal{A})\psi_\omega} = \mathcal{H}$ one finds that the central support of E_0 is one and consequently $\mathcal{R}$ is a von Neumann algebra of type I.

Now ψ_ω is a cyclic vector for $\pi(\mathcal{A})$ and therefore also for $\mathcal{R}$. From this one sees that ψ_ω is a cyclic vector for $E_0 \mathcal{R} E_0$ in $E_0 \mathcal{H}$. But an abelian algebra with cyclic vector is equal to its commutant, i.e. $(E_0 \mathcal{R} E_0)' = E_0 \mathcal{R} E_0$. Since ψ_ω is cyclic for $\mathcal{R}$ and since $E_0 \in \mathcal{R}$ has central support $\mathbb{1}$ one obtains from $E_0 \psi_\omega = \psi_\omega$ that the map $\mathcal{R}' \to \mathcal{R}' E_0$ is an isomorphism of von Neumann algebras. Consequently $\mathcal{R}'$ is abelian and is isomorphic to $E_0 \mathcal{R} E_0$. It remains to show that $\mathcal{R}'$ and $\mathcal{Z}_\pi$ coincide. Since E_0 is an abelian projection of central support 1 and since $E_0 \pi(\mathcal{A})'' E_0 = E_0 \mathcal{R} E_0$ and since E_0 commutes with $\mathcal{Z}_\pi$ it follows that $E_0 \mathcal{Z}_\pi$ and $E_0 \pi(\mathcal{A})'' E_0 = E_0 \mathcal{R} E_0$ coincide. Hence $E_0 \mathcal{Z}_\pi$ and $E_0 \mathcal{R}'$ are the same. This implies $\mathcal{Z}_\pi = \mathcal{R}'$. $\qquad\square$

After all these preparations we are able to prove a result which is called the additivity of the spectrum. But for the formulation and the proof of this result we need some notation and terminology:

IV.4.7 Definition:

Let $\{\mathcal{A}, \mathbb{R}^d, \alpha\}$ be a C^*-dynamical system and $\{\pi, \mathcal{H}, U(a)\}$ a covariant representation such that

(α) $U(a)$ is strongly continuous,

(β) $\mathcal{Z}_\pi$ commutes with $U(a)$.

Denote by $\operatorname{spec} U$ the spectrum of $U(a)$ and let B_r denote the ball of radius r. Then $p \in \operatorname{spec} U$ implies that $E(p + B_r) \neq 0$ for every $r > 0$ where E is the spectral projection of the group representation $U(a)$. By $F(p + B_r)$ we will denote the support of $E(p+B_r)$ in $\mathcal{Z}_\pi$. We will call points $p_1, \cdots, p_n \in \operatorname{spec} U$ **compatible** if for every $r > 0$ one has $\prod_{i=1}^{n} F(p_i + B_r) \neq 0$.

With these definitions one obtains

IV.4.8. Theorem:

Let $\{\mathcal{A}(O), \mathcal{A}, \mathbb{R}^d, \alpha\}$ be a local net, ω an invariant state, and let $\{\pi, \mathcal{H}, \psi_\omega, U(a)\}$ be the canonical representation defined by ω. Assume that $U(a)$ is a strongly continuous group representation which commutes with $\mathcal{Z}_\pi$.

If p_1, p_2 both belong to spec U *and if they are compatible then $p_1 + p_2 \in$* spec U *and the three points p_1, p_2, and $p_1 + p_2$ are compatible.*

Proof: Notice first that all points of spec U are compatible with each other if π is a factor representation. If π is not a factor representation and if p_1, p_2 are both elements of spec U then it can happen that there exists a central projection F such that $p_1 \in$ spec UF, $p_2 \notin$ spec UF. In such a situation nothing can be said about $p_1 + p_2$.

Denote by $\widehat{\mathcal{A}}_\pi$ the concrete C^*–algebra generated by all $\pi(\mathcal{A}(O))''$. Then the system $\{\pi(\mathcal{A}(O))'', \widehat{\mathcal{A}}_\pi, \mathbb{R}^d, \alpha\}$ defines again a local net and all that which has been proved before also holds for this local net. The strong continuity of $U(a)$ implies the weak continuity of $\pi(\alpha_a x)$ for every $x \in \widehat{\mathcal{A}}_\pi$. If now $f \in \mathcal{L}^1(\mathbb{R}^d)$ then the expression

$$\pi(x(f)) = \int \pi(\alpha_a x) f(a) \, da$$

is well-defined and belongs to $\widehat{\mathcal{A}}_\pi$.

If $p \in$ spec U and if supp $\mathcal{F}^{-1} f \subset p + B_r$ then one has $E(p + B_r) \pi(x(f)) \psi_\omega = \pi(x(f)) \psi_\omega$. Moreover, taking functions f the Fourier transforms of which are equal to one on the set $p + B_{r/2}$ it can be seen that the set of vectors

$$\{E(p + B_{r/2}) \pi(x(f)) \psi_\omega; \text{ supp } \mathcal{F}^{-1} f \subset p + B_r, \, x \in \mathcal{A}\}$$

is total in $E(p + B_{r/2}) \mathcal{H}$.

Assume that the statement of the theorem is not correct. Then there exists r_0 such that $F(p_1 + p_2 + B_r) F(p_1 + B_r) F(p_2 + B_r) = 0$ for $r \leq r_0$, and hence

$$E(p_1 + p_2 + B_r) F(p_1 + B_r) F(p_2 + B_r) = 0 \quad \text{for} \quad r \leq r_0.$$

Now take functions f_i, $i = 1, 2$ such that supp $\mathcal{F}^{-1} f_i \subset p_i + B_{r_0/2}$. Then

$$E(p_1 + p_2 + B_{r_0}) \pi(x_1(f_1)) \pi(x_1(f_2)) \psi_\omega = \pi(x_1(f_1)) \pi(x_2(f_2)) \psi_\omega.$$

Multiplying by $F(p_1 + B_{r_0}) F(p_2 + B_{r_0})$ one obtains zero and hence replacing x_2 by $\alpha_{\lambda a} x_2$ with a spacelike one finds

$$0 = (\psi_\omega, \pi(\alpha_{\lambda a} x_2(f_2)^*) \pi(x_1(f_1)^*)$$
$$\pi(x_1(f_1)) \pi(\alpha_{\lambda a} x_2(f_2)) F(p_1 + B_{r_0}) F(p_2 + B_{r_0}) \psi_\omega).$$

Taking the limit $\lambda \to \infty$ one obtains from Proposition IV.4.5

$$0 = (\psi_\omega, \pi(x_1(f_1)^*) \pi(x_1(f_1)) F(p_1 + B_{r_0/2}) E_0 F(p_2 + B_{r_0/2})$$
$$\pi(x_2(f_2)^*) \pi(x_2(f_2)) \psi_\omega)$$

for every $x_i \in \mathcal{A}$, $i = 1, 2$ and supp $\mathcal{F}^{-1} f_i \subset p_i + B_{r_0/2}$. Since all the operators are positive this relation remains true if we replace $F(p_i + B_{r_0})$ by a smaller operator, e.g. $F(p_i + B_{r_0/4})$. But by the choice of supp $\mathcal{F}^{-1} f_i$ we know that $\{F(p_i + B_{r_0/2}) E_0 \pi (x_i(f_i)^* x_i(f_i)) E_0\}$ are total in the set of positive operators in $E_0 F(p_i + B_{r_0/4}) \mathcal{Z}_\pi$. Therefore the right hand side of the last equation is not identically zero. This gives the desired contradiction and the theorem is proved. □

Finally we want to prove a characterization of vacuum states. This will require additivity of the spectrum proved in the last theorem.

IV.4.9 Definition:

Let $\{\mathcal{A}(O), \mathcal{A}, \mathbb{R}^d, \alpha\}$ be a theory of local observables. A state $\omega \in \mathcal{A}^*(\overline{V}^+)$ will be called a vacuum state if

(α) ω is an invariant state,
(β) the canonical representation $U(a)$ (which is automatically continuous since $\omega \in \mathcal{A}^*(\overline{V}^+)$) of the translation group fulfils the spectrum conditions.

With this notation the main result of this section is

IV.4.10 Theorem:

Let $\{\mathcal{A}(O), \mathcal{A}, \mathbb{R}^d, \alpha\}$ be a theory of local observables. Then every invariant state belonging to $\mathcal{A}^*(\overline{V}^+)$ is a vacuum state.

Proof: Let $\{\pi, \mathcal{H}, U(a)\}$ be the canonical representation defined by ω. Since $\omega \in \mathcal{A}^*(\overline{V}^+)$ there exists a different representation $V(a)$ of the group $\mathbb{R}^d$ which implements the same automorphism α_a, which belongs to $\pi(\mathcal{A})''$ and which fulfils the spectrum condition. But this shows that the center $\mathcal{Z}_\pi$ commutes with $V(a)$ and hence with $U(a)$. Since $V(a)$ and $U(a)$ commute we can write $U(a) = V(a) V'(a)$ with $V(a) \in \pi(\mathcal{A})''$ and $V'(a) \in \pi(\mathcal{A})'$. Let $\mathcal{R}$ again denote the von Neumann algebra generated by $\pi(\mathcal{A})$ and $\{U(a)\}$. Since $V'(a)$ is a commuting set of operators one sees that $V'(a)$ belongs to $\mathcal{R}'$. But from Proposition IV.4.6 one knows that $\mathcal{R}'$ coincides with $\mathcal{Z}_\pi$ and hence $V'(a)$ belongs to $\mathcal{Z}_\pi$. This means that $\mathcal{R}$ and $\pi(\mathcal{A})''$ coincide.

Let E_0 again be the projection onto all vectors in $\mathcal{H}$ which are invariant under $U(a)$. Then $E_0 V(a)$ and $E_0 V'(a)$ are inverses of each other (on $E_0 \mathcal{H}$). Since the spectrum of $V(a)$ belongs to V^+ it follows from the last relation that the spectrum of $V'(a)$ belongs to $-V^+$. Now assume that spec $V'(a)$ is different from zero hence the spectrum of $U(a)$ has points in $-V^+$. Assume that $p \neq 0$, $p \in -V^+$ is in the spectrum of $V'(a)$. Let $F(< p, 0 >)$ be the spectral projection of $V'(a)$ on the closed order interval $< p, 0 >$ (which is not zero by assumption). Then the spectrum of $U(a) F(< p, 0 >)$ is contained in $p + V^+$. Assume $q \neq 0$ and $q \in \operatorname{spec} U(a) F(< p, 0 >)$ such that $q \notin V^+$.

Then by Theorem IV.4.8 also $nq \in \operatorname{spec} U(a)F(< p, 0 >)$. However, this contradicts the statement $\operatorname{spec} U(a)F(< p, 0 >) \subset p + V^+$. This implies $\operatorname{spec} U(a)F(< p, 0 >) \subset V^+$. Since $F(< p_i, 0 >) \to 1$ for p_i such that $\cup < p_i, 0 >= -V^+$ we conclude that $\operatorname{spec} U(a) \subset V^+$. □

In the course of the proof of the Theorem we have proved something more, namely:

IV.4.11 Corollary:

Let $\{\mathcal{A}(O), \mathcal{A}, \mathbb{R}^d, \alpha\}$ be a theory of local observables and ω a vacuum state. Then:

(1) $\pi_\omega(\mathcal{A})''$ *is of Type I with* $\pi_\omega(\mathcal{A})' = \mathcal{Z}_\pi$.
(2) *The canonical representation $U(a)$ of the translation group coincides with the minimal representation described in* IV.2.
(3) *The spectrum of $U(a)$ is an additive set, which means if $p_1, p_2 \in \operatorname{spec} U$ and if p_1, p_2 are compatible then $p_1 + p_2 \in \operatorname{spec} U$.*

IV.5 Additivity properties of the spectrum

The investigation of the theory of local observables $\{\mathcal{A}(O), \mathcal{A}, \mathbb{R}^d, \alpha\}$ will be continued. Let $\omega \in \mathcal{A}^*(V^+)$ be a vacuum state and let $\{\pi_0, \mathcal{H}_0, U_0(a)\}$ be the canonical representation of the algebra $\mathcal{A}$ defined by ω. Here $U_0(a)$ denotes the canonical representation, which exists since ω is an invariant state, and which coincides with the unique minimal representation described in section VI.2. In case that π_0 is a factor representation we know from Corollary IV.4.11 that the spectrum of $U_0(a)$ is an additive set. We want to discuss the physical meaning of this result in order to learn what can be expected for arbitrary positive energy representations.

In section I.1 we have identified the representations fulfilling the spectrum condition with the representations of particle physics. It is customary to classify the different particles by their charge, their mass, and their spin. In standard quantum field theory there exists a field to every particle. If the particle is charged then the field must also be charged. Hence this is not an observable quantity. Therefore it does not appear in our theory, but nevertheless the charged fields can be used as a guidance.

In particle physics factor representations are also called superselection sectors because a superposition of vectors in two different sectors does not lead to observable consequences. Therefore the charges describing the different sectors must commute with every observable. Since the time development belongs to the weak closure of the observable algebra it acts separately in every sector. If for example one considers a theory with only one charged field besides the observable algebra, then a charged field leads from one representation of the observable algebra to an inequivalent one. If we use products of this charged field then all sectors are obtained by applying these products to the vacuum sector. This picture can be generalized to several charged fields. In such a situation one can characterize the factor representations of the observable algebra by charge quantum numbers. Now we can associate the quantum number Q of the sector to every vector in a charged sector. Physically we identify most of these vectors with a family of particles which together must carry the charge Q. In particular in the vacuum sector a vector represents particles with no charge. Since the union of two families of particles with no charge is again a family with no charge, to every two vectors in the vacuum sector there exists a third one representing the union. We do not know which vector this will be, but we can say that the corresponding four-momenta should be additive. This is just the additivity of the spectrum in the vacuum sector. If we have a charged particle and add this to a family of particles then the charge changes. This implies that we obtain another sector. Therefore we remain in the same sector only if we add a group of particles carrying the charge zero. If S_Q is the spectrum of the translations in the sector Q and if S_0 is that one in the vacuum sector, then we expect:

$$S_Q + S_0 \subset S_Q.$$

This argument uses the spectra of the translations in two different sectors. We are interested in a statement concerning only one sector. If the above picture with the charged fields is correct, then the adjoint of such a field carries the opposite charge. If a sector is obtained by applying a charged field to the vacuum sector, then one must find a sector with opposite charge to each sector. Since the spectrum does not change by this charge conjugation the following holds:

$$S_{-Q} = S_Q.$$

Moreover, a particle carrying charge Q and its antiparticle result in a neutral pair. Hence we expect:

$$S_0 \supset S_Q + S_{-Q} = S_Q + S_Q.$$

If we insert this relation into the above equation we obtain

$$S_Q + S_Q + S_Q \subset S_Q.$$

I am convinced that one day it will turn out that this is a correct equation. Moreover, I believe that it can be proved by constructing an envelope of holomorphy for one edge of the wedge problem. However, the techniques we have provided in chapter III are only sufficient to prove a weaker statement.

IV.5.1 Theorem:

Let $\{\pi, \mathcal{H}, U(a)\}$ be a factor representation of a theory of local observables fulfilling the spectrum condition, and assume that $U(a)$ is the unique minimal representation of the translations. Assume the spectrum of $U(a)$ consists of two parts:

(a) *Isolated hyperboloids with the masses $m_0 < m_1 \cdots m_i < \cdots$.*
(b) *The rest starting at $m_c > m_i$ where m_c denotes the beginning of the continuous spectrum or the first accumulation point of the $m_i's$.*

Then

$$3m_0 \geq m_c.$$

Proof: We use a technique similar to that described in IV.1. Define the functions

$$F_{x,\psi}^{0,+}(a) = (\psi, \pi(x^*)U(a)\pi(x)\psi)$$
$$F_{x,\psi}^{0,-}(a) = (\psi, U(a)\pi(x)U(-a)\pi(x^*)U(a)\psi).$$

Choose $\epsilon > 0$ and let $m_0, m_1, \cdots, m_i$ be the masses smaller than $m_c - \epsilon$. The functions F^+ and F^- are defined by

$$F_{x,\psi}^+(a) = (\Box - m_0^2)(\Box - m_1^2) \cdots (\Box - m_i^2)F_{x,\psi}^{0,+}(a)$$
$$F_{x,\psi}^-(a) = (\Box - m_0^2)(\Box - m_1^2) \cdots (\Box - m_i^2)F_{x,\psi}^{0,-}(a).$$

Furthermore we put

$$F_{x,\psi}(a) = F_{x,\psi}^+(a) - F_{x,\psi}^-(a).$$

Now let t be a vector in the interior of V^+ and x an element of $\mathcal{A}(D_t)$. Due to locality we obtain $F_{x,\psi}(a) = 0$ for $a \in D'_{2t}$. This implies that we can write

$$F_{x,\psi}(a) = G_{x,\psi}^+(a) - G_{x,\psi}^-(a)$$

with

$$\text{supp}\, G_{x,\psi}^+(a) \subset -2t + V^+$$
$$\text{supp}\, G_{x,\psi}^-(a) \subset 2t - V^+.$$

Taking now the Fourier transformation one sees that $\tilde{G}^+(p)$ is the boundary value of an analytic function holomorphic in $T^+ = \{z \in \mathbb{C}^d; \text{Im}\, z \in V^+\}$. Accordingly $\tilde{G}^-$ is the boundary value of an analytic function holomorphic in $T^- = -T^+$. On the real points one has

$$\tilde{G}^+(p) = \tilde{G}^-(p) \quad \text{for} \quad p \in \Gamma$$

where

$$\Gamma = \mathbb{R}^d \setminus \text{supp}\, \tilde{F}_{x,\psi}(p).$$

In order to determine the support of $\widetilde{F}_{x,\psi}(p)$ we must first fix the vector ψ. Let s be a vector in V^+ with $s^2 = 1$ and let $\delta > 0$. Then the spectral projection $E(D_{0,(m_0+\delta)s})$ is not zero and hence there exist vectors ψ which are eigenvectors of this projection. Let now $V_m^+ = \{x \in V^+; x^2 \geq m^2\}$. Then we have $\operatorname{supp} \widetilde{F}_{x,\psi}^{0,+}(p) \subset V_{m_0}^+$ and hence $\operatorname{supp} \widetilde{F}_{x,\psi}^+(p) \subset V_{m_c-\epsilon}^+$. Since the function $\widetilde{F}_{x,\psi}^-(p)$ contains a shift of the spectrum induced by the vector ψ it follows that the wave operators do not affect the support of the functions. This means that the support of $\widetilde{F}_{x,\psi}^{0,-}(p)$ and $\widetilde{F}_{x,\psi}^-(p)$ coincide. Hence we obtain:

$$\operatorname{supp} \widetilde{F}_{x,\psi}^-(p) \subset 2(m_0 + \delta)s - V_{m_0}^+.$$

The two together amount to

$$\operatorname{supp} \widetilde{F}_{x,\psi}(p) \subset V_{m_c-\epsilon}^+ \cup \{2(m_0 + \delta)s - V_{m_0}^+\}.$$

Now we apply the J.-L.-D. representation described in III.4 to the coincidence domain Γ. This means that we have to consider all hyperboloids with apex in the order interval $D_{0,2(m_0+\delta)s}$ which do not enter Γ. These are the hyperboloids with mass $\mu = m_c - 2m_0 - 2\delta - \epsilon$. This implies that the support of $\widetilde{F}_{x,\psi}(p)$ remains stable under the J.-L.-D. construction if $\mu \leq m_0$. If $\mu \geq m_0$ then we obtain an enlargement of the coincidence domain Γ which implies that in this case $\widetilde{F}_{x,\psi}^-(p)$ has its support in $2(m_0 + \delta)s - V_\mu^+$.

The condition $\mu \leq m_0$ means $m_c - 2m_0 - 2\delta - \epsilon \leq m_0$ or $m_c - 2\delta - \epsilon \leq 3m_0$. If this condition holds for arbitrarily small ϵ and δ then we find $m_c \leq 3m_0$. This is the inequality claimed in the Theorem.

Next assume that $m_c > 3m_0$. Then there exist $\epsilon > 0$ and $\delta > 0$ such that $\mu > m_0$. In this situation we obtain

$$\operatorname{supp} \widetilde{F}_{x,\psi}^-(p) \subset \{2(m_0 + \delta)s - V_\mu^+\},$$

since the supports of $\widetilde{F}_{x,\psi}^+(p)$ and $\widetilde{F}_{x,\psi}^-(p)$ have empty intersection. Therefore the theorem is proved if we can show that this is in contradiction to the assumption that the spectrum of $U(a)$ starts at m_0. This will be done in a separate Lemma:

IV.5.2 Lemma:

Keep the notation of Theorem IV.5.1. The following two assumptions contradict each other.

i) *The spectrum of $U(a)$ starts exactly at m_0.*

ii) *For every ψ with $\operatorname{supp} \psi$ contained in the order interval $< 0, (m_0 + \delta)s >$, $s \in V^+$, $s^2 = 1$, $\delta > 0$, and every $x \in \mathcal{A}(O)$ for some bounded region O, the functions $\widetilde{F}_{x,\psi}^-(p)$ have their support in $\{2(m_0 + \delta)s - V_\mu^+\}$, where $\mu > (m_0 + 2\delta)$ is a fixed number.*

Proof: Since we obtain the support of $\widetilde{F}^{0,-}_{x,\psi}(p)$ by convoluting $-\operatorname{spec} U(a)$ twice with the support of ψ we see that multiplication with $(p^2 - m_j^2)$ does not change the support. Hence the support of $\widetilde{F}^{0,-}_{x,\psi}(p)$ is the same as that of $\widetilde{F}^{-}_{x,\psi}(p)$.

Now choose a compact set Γ outside of $\{2(m_0+\delta)s - V_\mu^+\}$ and a function with compact support $\tilde{f}(p) \in \mathcal{D}$ with $0 \le \tilde{f}(p) \le 1$, $\tilde{f}(p) = 0$ for $p \in 2(m_0 + \delta)s - V_\mu^+$, and $\tilde{f}(p) = 1$ for $p \in \Gamma$. Then

$$\int \widetilde{F}^{0,-}_{x,\psi}(p)\tilde{f}(p)\mathrm{d}p = 0 = \int F^{0,-}_{x,\psi}(a)f(-a)\mathrm{d}a.$$

Next choose ϵ such that $16\epsilon\|f\|_1 < 1$, and a compact K such that $\int_{\mathbb{R}^d\setminus K} |f(a)|\mathrm{d}a \le \epsilon\|f\|_1$. With this we obtain:

$$\left| \int_K (\psi, U(a)xU(-a)x^*U(a)\psi)f(-a)\mathrm{d}a \right| \le \epsilon\|f\|_1\|x\|^2\|\psi\|^2. \qquad (*)$$

Let $q \in \operatorname{spec} U \cap D_{(0,(m_0+\delta)s)}$ and let E denote the spectral projections of $U(a)$. Now choose a small sphere Δ around q. By applying Schwarz' lemma, we learn that we can choose the sphere so small that

$$\|(U(a) - e^{i(q,a)}E(\Delta)\| < \epsilon \quad \text{for} \quad q \in K$$

holds. Inserting this into (*) we obtain

$$\left| \int_K (\psi, xU(-a)x\psi)e^{2i(q,a)}f(-a)\mathrm{d}a \right| \le 3\epsilon\|f\|_1\|x\|^2\|\psi\|^2$$

and hence

$$\left| \int_{\mathbb{R}^d} (\psi, xU(-a)x\psi)e^{2i(q,a)}f(-a)\mathrm{d}a \right| \le 4\epsilon\|f\|_1\|x\|^2\|\psi\|^2. \qquad (**)$$

Since $\tilde{f}(p) \ge 0$ and since $\tilde{f}(p) = 1$ for $p \in \Gamma$ one has:

$$E(-\Gamma + 2q) \le \int U(-a)f(-a)e^{2i(q,a)}\mathrm{d}a.$$

Inserting this into (**) one has

$$(x^*\psi, E(-\Gamma + 2q)x\psi) \le 4\epsilon\|f\|_1\|x\|^2\|\psi\|^2. \qquad (***)$$

Remember that we are dealing with a factor representation. If $E(-\Gamma+2q)$ and $E(\Delta)$ are both non-zero then either $E(\Delta)$ is equivalent to a subprojection on $E(-\Gamma + 2q)$ or vice versa. This implies that there exists a partial isometry V in π'' with domain in $E(\Delta)$ and range in $E(-\Gamma + 2q)$. Let us choose the vector ψ in the domain of V. Then by Kaplansky's density theorem there exists an operator $y \in \pi(\mathcal{A})$ with $\|y\| = 1$ and $\|(y-V)\psi\| < \epsilon_1\|\psi\|$.

Moreover, there is an operator x^* in $\mathcal{A}(O)$ for some O with $\|x^* - y\| \leq \epsilon_2$. By choice of V we have $(V\psi, E(-\Gamma + 2q)V\psi) = \|\psi\|^2$ which implies:

$$(x^*\psi, E(-\Gamma + 2q)x^*\psi) \geq (1 - (1 + \epsilon_2)(\epsilon_1 + \epsilon_2)).$$

On the other hand we obtain for this x^*, by equation $(***)$, the estimate

$$(x^*\psi, E(-\Gamma + 2q)x^*\psi) \leq 4\epsilon \|f\|_1 (1 + \epsilon_2)^2.$$

These two inequalities contradict each other provided we choose all the ϵ in such a way that

$$4\epsilon \|f\|(1 + \epsilon_2)^2 < (1 - (1 - \epsilon_2)(\epsilon_1 + \epsilon_2))$$

holds. Hence the projection $E(-\Gamma + 2q)$ is zero. Now we choose for q the vector $m_0 s$ and for Γ a small sphere around the point $(2m_0 + \delta - (m_0 + 2\delta))s$. This sphere exists since $\mu > m_0 + 2\delta$. In this case $-\Gamma + 2q$ is a small sphere around $2m_0 s - [(2m_0 + \delta)s - (m_0 + 2\delta)s] = m_0 s$. This contradicts the assumption that the spectrum of $U(a)$ starts at m_0. □

IV.6 Absence of classical fields

Looking at the axioms for theories of local observables one sees that there is one trivial way of fulfilling them. Namely we take an abelian algebra $\mathcal{A}$ and identify every $\mathcal{A}(O)$ with $\mathcal{A}$. Furthermore, if we choose α_a to be trivial then of course it is implemented by $U(a) = 1$ which has its spectrum at zero.

This example raises the question whether or not one can have classical field theories fulfilling all assumptions of chapter I. Here we will identify classical field theories with abelian algebras. It is well known that classical field theories exist as long as one does not require the spectrum condition. It will be the content of the following result that there exist no classical fields fulfilling the spectrum condition. But first some terminology:

IV.6.1 Definition:

Let $\{\mathcal{A}(O), \mathcal{A}, \mathbb{R}^d, \alpha\}$ be a theory of local observables. Then the theory $\{\mathcal{B}(O), \mathcal{B}, \mathbb{R}^d, \alpha, \overline{V}^+\}$ is called a sub-theory if

(i) *$\mathcal{B}(O)$ is a sub-C^*-algebra of $\mathcal{A}(O)$.*
(ii) *The automorphisms $\alpha^{\mathcal{B}}$ coincide with the restriction of $\alpha^{\mathcal{A}}$ to the subalgebra $\mathcal{B} \subset \mathcal{A}$, which is an invariant subalgebra by this construction.*

$\mathcal{A}$ allows a faithful representation $\{\pi, U(a), \mathcal{H}\}$, by assumption. For $\mathcal{B}$ we obtain a faithful representation by restriction.

IV.6.2 Theorem:

Let $\{\mathcal{B}(O), \mathcal{B}, \mathbb{R}^d, \alpha\}$ be a sub-theory of the theory of local observables $\{\mathcal{A}(O), \mathcal{A}, \mathbb{R}^d, \alpha\}$, and assume that $\mathcal{B}$ is an abelian C^*-algebra. Then

(1) The elements of $\mathcal{B}$ are translationally invariant, i.e. $y \in \mathcal{B}$ and $a \in \mathbb{R}^d$ implies $\alpha_a(y) = y$.

(2) $\mathcal{B}$ is a subset of the center of $\mathcal{A}$, $\mathcal{B} \subset \mathcal{Z}(\mathcal{A})$.

Proof: Let $F(\overline{V}^+)$ be the projection in $\mathcal{B}^{**}$ defined in II.7.1. Then α_a is an inner automorphism of $\mathcal{B}^{**} F(\overline{V}^+)$ and hence trivial since $\mathcal{B}$ is abelian. By the remark following Definition IV.6.1 we know that $\mathcal{B}F(\overline{V}^+)$ is faithful, and hence α_a acts trivially on $\mathcal{B}$. If $y \in \mathcal{B}(O)$ then it commutes with $\mathcal{A}(O_1)$ if $O_1 \subset O'$. But since y is translation invariant it commutes with $\mathcal{A}(O)$ for every O and hence $y \in \mathcal{Z}(\mathcal{A})$. Since those elements are norm-dense in $\mathcal{B}$ one has $\mathcal{B} \subset \mathcal{Z}(\mathcal{A})$. □

Another result concerning local observables is about localization. If the interpretation of the observables given in Section I.1 is correct then it should be impossible to localize an observable at a point. An exception are of course trivial observables like the identity or elements of the center of $\mathcal{A}$.

IV.6.3 Theorem:

Let $\{\mathcal{A}(O), \mathcal{A}, \mathbb{R}^d, \alpha\}$ be a theory of local observables. For a point $a \in \mathbb{R}^d$ define

$$\mathcal{A}(a) = \bigcap_{a \in O} \mathcal{A}(O),$$

then $\mathcal{A}(a) \subset \mathcal{Z}(\mathcal{A})$.

Proof: By the definition of the algebra at a point one has $\alpha_a \mathcal{A}(b) = \mathcal{A}(a + b)$. Hence the algebra $\mathcal{B}$ generated by $\{\mathcal{A}(a), \ a \in \mathbb{R}^d\}$ defines an invariant subalgebra of $\mathcal{A}$. If one can show that α_a acts trivially on $\mathcal{B}$ then one sees from the definition of $\mathcal{B}$ that it is an abelian algebra, and the result is implied by Theorem IV.6.2. Let $V(a)$ be the unique minimal representation of the translations in $\mathcal{B}^{**} F(\overline{V}^+)$ implementing α_a on $\mathcal{B}$. If we show that the spectrum of $V(a)$ is the point zero then it will follow that α_a acts trivially on $\mathcal{B}$ and hence $\mathcal{B}$ is abelian.

To this end let ψ be a vector with support in the order interval $< 0, p >$, $p \in V^+$ and let G be the central support of $F(< 0, p >)$. Then we investigate $V(a) G$. Let $x \in \mathcal{A}(\{0\})$, $F^+ = (\psi, x^* V(a) x \psi)$ and $F^- = (\psi, V(a) x V(-a) x^* V(a) \psi)$. From the continuity of the representation one has $F^+(a) = F^-(a)$ for $a^2 \le 0$. Let $F(a) = F^+(a) - F^-(a)$. $F(a)$ is zero for $a^2 \le 0$ and it is bounded and continuous elsewhere.

Fix a coordinate system such that the vector p becomes $(p, 0)$. Then one obtains

$$\int F(a)\delta(a_0 - a_1)f(\mathbf{a})\mathrm{d}\mathbf{a} = 0$$

for every $f \in \mathcal{L}^1(\mathbb{R}^{d-1})$ and every chosen 1-direction. Writing the vector $q = (q_0, q_1, \widehat{q})$ where $\widehat{q}$ represents the other variables one obtains in momentum space

$$\int \widetilde{F}(q)\widetilde{f}(-q_0 + q_1, -\widehat{q})\,\mathrm{d}q = 0$$

where the tilde indicates the inverse Fourier transformation. Note that $\widetilde{F}(q)$ and $\mathcal{F}^{-1}F^+$ coincide in $\overline{V}^+ \setminus \{2p + -\overline{V}^+\}$ and that supp $\widetilde{F}(q) \subset \{\overline{V}^+ \cup (2p - \overline{V}^+)\}$. Choosing f such that $\widetilde{f}(-q_0 + q_1, \widehat{q}) = 0$ for $-q_0 + q_1 \geq -2p_0$ we obtain

$$\int \widetilde{F}(q)\widetilde{f}(-q_0 + q_1, \widehat{q})\,\mathrm{d}q = \int \mathcal{F}^{-1}\widetilde{f}(-q_0 + q_1, \widehat{q})\,\mathrm{d}q.$$

Note that $(\mathcal{F}^{-1}F^+)(q)$ is a positive measure. Choosing $\widetilde{f} \geq 0$ we see that $(\mathcal{F}^{-1}F^+)(q) = 0$ for $-q_0 + q_1 < -2p_0$. Changing the direction of q_1 we have

$$\text{supp}\,\mathcal{F}^{-1}F^+ \subset \{\overline{V}^+ \cap (2p - \overline{V}^+)\}.$$

From this one sees that F^+ is entire analytic which implies that

$$(\psi, x^*V(a)x\psi) = (\psi, V(a)xV(-a)x^*V(a)\psi)$$

holds for arbitrary a. Since the vectors with compact support in momentum space are dense one has

$$[x^*, \alpha_a(x)] = 0 \quad \text{for all} \quad a \in \mathbb{R}^d, \ x \in \mathcal{A}(\{0\}).$$

Using polarization one obtains:

$$[y, \alpha_a(x)] = 0 \quad \text{for all} \quad a \in \mathbb{R}^d \ x, y \in \mathcal{A}(\{0\}).$$

From this one concludes that $\mathcal{B}$ is an abelian algebra. $\qquad\qquad \square$

IV.7   Notes and remarks

(1) The edge of the wedge theorem was invented in order to prove dispersion relations for scattering amplitudes [BOT]. See also K. Symanzik [Sy]. The first person to use a representation for the commutator was H. Lehmann [Le].

(2) The concept of minimal translation was introduced by D. Olesen and G.K. Pedersen [OP] in the case of a one-parameter group. The generalization to the higher dimensional case does not work under general circumstances.

Special conditions are necessary. The first construction of minimal representation was given by Borchers and Buchholz [BB]. However, this method was somewhat indirect. The first direct proof was presented by the author [Bch87].

(3) As we have seen, the representation of the translation group is automatically the minimal one if it is generated from a vacuum state. This led to the observation by Borchers [Bch65] that in every sector which is either the vacuum sector, or which is generated from the vacuum by a charged field, the spectrum of the translations has a support which is a Lorentz invariant set. In the same paper it was shown that the Lebesgue continuous and the discontinuous part of the spectrum also have Lorentz invariant sets as supports. For the general situation Borchers and Buchholz showed in [BB] that the lower boundary is invariant under Lorentz transformations. The general invariance property of the spectral support was obtained in [Bch85].

(4) The first proof of the cluster property is due to the author [Bch62]. A systematic study of this property was started by Doplicher, Kadison, Kastler, and Robinson [DKKR] using the notation of asymptotic Abelian systems introduced by Doplicher, Kastler, and Robinson in [DKR] and independently by Ruelle [Ru]. This notion has been weakend by Lanford and Ruelle [LR] introducing the concept of G-Abelian systems. The most general concept leading to the cluster property has been introduced by Størmer [Stø]. He called it large groups of automorphisms. One important consequence of the cluster property of the vacuum state is the additivity of the spectrum. The result is due to Wightman [Wi64]. If one has a theory with charged fields then the charged sectors can be labeled by $\mathbf{Z}^n$. If we denote the spectrum of the translations in the sector $\mathbf{h}$ by $S_{\mathbf{h}}$, then Wightman's result generalizes to $S_{\mathbf{h}_1} + S_{\mathbf{h}_2} \subset S_{\mathbf{h}_1 + \mathbf{h}_2}$ [Bch65].

(5) The proof of the general additivity property of the spectrum is due to the author [Bch86]. In my opinion the result presented here is not the best one. I conjecture that the relation $3S \subset S$ must hold for every factor representation obeying spectrum condition. However, at present there are not enough techniques available to prove it.

There is one more result which points in the same direction. It states that the spectrum cannot be finite in the mass. The proof of this result uses slightly different techniques from those discussed in this note for constructing the envelope of holomorphy. The result is due to Armbrüster [Arm].

As promised in the beginning of this chapter we want to give examples showing the independence of the locality and the spectrum condition.

The spectrum condition is necessary for the particle interpretation. But in order to obtain statements about the spectrum we also had to use the locality condition. The first example shows that it is not sufficient to replace locality by asymptotic locality.

Example: Let $\Delta \subset \overline{V^+}$ and μ be a Lebesgue continuous positive measure on Δ. Set $\mathcal{H} = \mathcal{L}^2(\Delta, \mu)$. Then on $\mathcal{H}$ exists a continuous unitary representation of the translations which fulfills the spectrum condition. This is given by $U(a)f(p) = e^{i(a,p)}f(p)$.

Define $\mathcal{A}$ to be the algebra of compact operators on $\mathcal{H}$. Using the Riemann–Lebesgue lemma one easily sees that this system is asymptotically abelian.

How to construct an example satisfying the locality condition but not the spectrum condition is indicated by Theorem IV.6.2.

Example: Let $\mathcal{H} = \mathcal{L}^2(\mathbb{R}^d)$ with respect to the Lebesgue measure. The map $f(b) \longrightarrow f(b - a)$ defines a continuous unitary representation of the translations. Taking for $\mathcal{A}$ the abelian algebra of bounded continuous functions on $\mathbb{R}^d$ we obtain a system on which the translations do not act trivially. By Theorem IV.6.2 this algebra has no representation fulfilling the spectrum condition (except every operator is mapped onto a multiple of 1).

From this example one can construct more complicated ones for istance combining it with a free field. If the space is two dimensional then there exists an example of different nature constructed by Doplicher, Regge and Singer [DRS].

Bibliography

General references

[[BT]]   Behnke H. und Thullen P. (1970): *Theorie der Funktionen mehrerer kom-plexer Veränderlicher* , Ergebnisse der Mathematik und ihrer Grenzgebie-te Bd **51**, Springer Berlin, Heidelberg, New York.

[[BM]]   Bochner S. and Martin W.T. (1948): *Several complex Variables*, Princeton University Press.

[[CH]]   Courant R. and Hilbert D. (1962): *Methods of Mathematical Physics* Vol. II, Interscience Publishers, Inc. New York, London.

[[Em]]   Emch G.G. (1972): *Algebraic Mehods in Statistical Mechanics and Quan-tum Fields*, Wiley-Interscience, New-York, London, Sidney, Toronto.

[[Fri]]   Friedrichs K.O. (1953): *Mathematical aspects of the quantum theory of fields* , Interscience Publishers, Inc. New York.

[[Gui]]   Guichardet A. (1968): *Algèbres d'observables associées aux relations de commutation* , Librairie Armand Colin, Paris.

[[GR]]   Gunning R.C. and Rossi H. (1965): *Analytic Functions of Several Complex Variables* , Prentice-Hall, Inc. Englewood Cliffs N.J.

[[Haa]]   Haag R. (1992): *Local Quantum Physics*, Springer Verlag, Berlin-Heidelberg-New York.

[[Hör]]   Hörmander L. (1966): *An introduction to complex analysis in several vari-ables* , D. van Nostrand Co. inc. Princeton, New York, London.

[[JR]]   Jauch J.M. and Rohrlich F. (1955): *The Theory of Photons and Electrons*, Addison-Wesley Publishing Co, Inc. Cambridge Mass.

[[Neu]]   Neumann J. von (1932): *Mathematische Grundlagen der Quantenmecha-nik*, Springer, Berlin.

[[PW]]   Paley R. and Wiener N. (1934): *Fourier Transforms in the Complex Do-main*, Amer. Math. Soc., New York.

[[Ped]]   Pedersen G.K. (1979): *C*-Algebras and their Automorphism Groups*, Aca-demic Press, London, New York, San Francisco.

[[RS]]   Reed M. and Simon B. (1975): *Methods of Modern Mathematical Physics.*, Vol II: *Fourier Analysis and Self-Adjoitness*, Academic Press; New York, San Francisco, London.

[[Rei]]   Reichenbach H. (1949): *Philosophische Grundlagen der Quantenmechanik*, Birkhäuser, Basel.

[[Schw]]   Schwartz L. (1957/59): *Théorie des Distribution* , Vol I (1957), Vol II (1959), Hermann, Paris.

[[Ta70]]   Takesaki M. (1970): *Tomita's Theory of Modular Hilbert Algebras and its Applications*, Lecture Notes in Mathematics, Vol. **128**, Springer-Verlag Berlin, Heidelberg, New York.

[[Ta79]]   Takesaki M. (1979): *Theory of Operator Algebras I*. Springer Verlag, Berlin-Heidelberg-New York.

[[Vla]]   Vladimirov V.S. (1964): *Methods of the Theory of Functions of Many Complex Variables* , The M.I.T. Press, Cambridge Mass, London (1966), translated from russian Nauka Press, Moscow (1964).

[[Var]]   Varadarajan V.S. (1968): *Geometry and Quantization Theory*, Van Norstrand, Princeton.

References

[AY]   Alcántara-Bode J. and Yngvason J. (1988): *Algebraic quantum field theory and noncommutative moment problems I: C^*-norms on the test function algebra modulo locality ideal*, Ann. Inst, Henri Poincaré **48**, 147-159.

[Ar61]   Araki H. (1961/62): *Einführung in die axiomatische Quantenfeldtheorie*, Vorlesungen an der E.T.H. Zürich, I. W.S. 1961/62 ausgearbeitet von K. Hepp u. F. Riahi, II. S.S. 1962 ausgearbeitet von W. Wyss.

[Ar64]   Araki H. (1964): *On the algebra of all local observables*, Progr. Theor. Phys. **32**, 844-854.

[AWo]   Araki H. and Woods E.J. (1966): *Complete Boolean algebras of type I factors*, Publ. Res. Inst. f. Math. Sci. Kyoto Univ. Series A, **2**, 157-242.

[AWy]   Araki H. and Wyss W. (1964): *Representations of Canonical Anticommutation Relations*, Helv. Physica Acta **37**, 136-159.

[Arm]   Armbrüster U. (1993): *Konstruktion spezieller Holomorphiehüllen*, Diplomarbeit Göttingen.

[Arv]   Arveson W.B. (1974): *On groups of automorphisms of operator algebras*, J. Functional Anal. **15**, 217-243.

[Asg]   Asgeirsson L. (1936): *Über eine Mittelwereigenschaft von Lösungen homogener linearer partieller Differentialgleichungen 2. Ordnung mit konstanten Koeffizienten*, Math. Ann. **113**, 321.

[BN]   Birkhoff G. and Neumann J. von (1936): *The Logic of Quantum Mechanics*, Ann. Math. **37**, 823-843.

[Bch61]   Borchers H.J. (1961): *Über die Vollständigkeit lorentzinvarianter Felder in einer zeitartigen Röhre*, Nuovo Cimento **19**, 787-796.

[Bch62]   Borchers H.J. (1962): *On the structure of the algebra of field operators*, Nuovo Cimento **24**, 214-236.

[Bch65]   Borchers H.J. (1965): *Local Rings and the Connection of Spin with Statistics*, Commun. Math. Phys. **1**, 281-307.

[Bch66]   Borchers H.J. (1966): *Energy and momentum as observables in quantum field theory*, Commun. Math. Phys. **2**, 49-54.

[Bch69]   Borchers H.J. (1969): *On the implementability of automorphism groups*, Commun. Math. Phys. **14**, 305-314.

[Bch70a] Borchers H.J. (1970): *On Groups of Automorphisms with Semi-bounded Spectrum*, in: Systèmes a un Nombre Infini de Degrés de Liberté, p. 125-142, Editions du C.N.R.S. Paris.

[Bch70b] Borchers H.J. (1970): *Strongly continuous automorphism groups on C^*-algebras*, in: 1968 Cargèse lectures in physics Vol 4 D. Kastler ed. p 31-59, Gordon and Breach, New York, London, Paris.

[Bch73a] Borchers H.J. (1973): *Über C^*-Algebren mit Lokalkompakten Symmetriegruppen*, Nachr. d. Akad. d. Wissensch. in Göttingen Heft 1, 1-18.

[Bch73b] Borchers H.J. (1973): *Über Ableitungen von C^*-Algebren*, Nachr. d. Göttinger Akad. der Wissensch. No 2 p. 1-18.

[Bch83]   Borchers H.J. (1983): *C^*-Algebras and Automophism Groups*, Commun. Math. Phys. **88**, 95-103.

[Bch84]   Borchers H.J. (1984): *Translation Group and Spectrum Condition*, Commun. Math. Phys. **96**, 1-13.

[Bch85]   Borchers H.J. (1985): *Locality and covariance of the spectrum*, Fizika **17**, 289-304.

122

[Bch86] Borchers H.J. (1986): *A Remark on Antiparticles* , in Lecture Notes in Physics Vol **257**, p. 268-280.

[Bch87] Borchers H.J. (1987): *On the Interplay between Spectrum Condition and Locality in Quantum Field Theory* , Contemporary Mathematics **62**, 143-152.

[Bch93] Borchers H.J. (1993): *Symmetry Groups of C^*-Algebras and Associated von Neumann Algebras*, in: Dynamics of Complex and Irregular Systems, p. 12-21, World Scientific Publishers.

[BB] Borchers H.J. and Buchholz D. (1985): *The Energy-Momentum Spectrum in Local Field Theories with Broken Lorentz-Symmetry*, Commun. Math. Phys. **97**, 169-185.

[BY] Borchers H.J. and Yngvason J. (1992): *From Quantum Fields to Local von Neumann Algebras*, Rev. Math. Phys., Special Issue, 15- 47.

[BZ] Borchers H.J. and Zimmermann W. (1964): *On the self-adjointness of field operators*, Il Nuovo Cimento **31**, 1047-1059.

[BJ] Born M. und Jordan P. (1925): *Zur Quantenmechanik* , Z. f. Phys. **34**, 858-888.

[Bre54] Bremermann H.J. (1954): *Über die Äquivalenz der pseudokonvexen Gebiete und der Holomorphiegebiete im Raum n komplexer Veränderlicher*, Math. Ann **128**, 63-91.

[Bre56] Bremermann H.J. (1956): *On the conjecture of the equivalence of plurisubharmonic functions and the Hartogs function* , Math. Ann. **131**, 76-86.

[BOT] Brehmermann H.J., Oehme R. and Taylor J.G. (1958): *Proof of dispersion relation in quantized field theories* , Phys. Rev. **109**, 2178-2190.

[Bros] Bros J. (1962): *Les Problèmes de Construction d'Enveloppes d'Holomorphie en Théorie Quantique des Champs*, Seminair Lelong no **4**.

[BEG] Bros J., Epstein H. and Glaser V. (1967): *On the connection between analyticity and Lorentz covariance of Wightman functions*, Commun. Math. Phys. **6**, 77-100.

[BMS] Bros J., Messiah A. and Stora R. (1961): *A problem of analytic completion related to the Jost-Lehmann-Dyson formula*, J. Math. Phys. **2**, 639-651.

[Brow] Browder F. (1963): *On the "Edge of the Wedge" Theorem* , Canad. J. Math. **15**, 125-131.

[Buch] Buchholz D. (1982/86): *The Physical State Space of Quantum Electrodynamics*, Commun.Math.Phys. **85**, 49-71 (1982), *Gauss' Law and the Infrared Problem*, Phys.Letters **174B**, 331-334 (1986).

[CT] Cartan H. and Thullen P. (1932): *Zur Theorie der Singularitäten der Funktionen mehrerer komplexer Veränderlicher* , Math. Ann. **106**, 617-647.

[Dir27] Dirac P.A.M. (1927): *The quantum theory of emission and absorption of radiation*, Proc. Roy. Soc London, Series A, **114** 243-265.

[Dir28] Dirac P.A.M. (1928): *The Quantum Theory of the Electron*, Proc. Roy. Soc. London, A, **117**, 610-624.

[Dop] Doplicher S. (1965): *An algebraic spectrum condition* , Commun. Math. Phys. **1**, 1-5.

[DKKR] Doplicher S., Kadison R.V., Kasler D. and Robinson D.W. (1967): *Asymptotically abelian systems* , Commun. Math. Phys. **6**, 101-120.

[DKR] Doplicher S., Kastler D. and Robinson D.W. (1966): *Covariance algebras in field theory and statistical mechnics* , Commun. Math. Phys. **3**, 1-28.

[DRS] Doplicher S., Regge T. and Singer I.M. (1968): *A geometrical model showing the independence of locality and positivity of the enery* , Commun. Math. Phys. **7**, 51-54.

[Dy] Dyson F.J. (1958): *Integral representation of causal commutators* , Phys. Rev. **110**, 1460-1464.

[Ep60] Epstein H. (1960): *Generalization of the "Edge of the Wedge" theorem*, Journ. Math. Phys. **1**, 524-531.

[Ep65] Epstein H. (1965): *Some analytic properties of scattering amplitudes in quantum field theory* , in: Particle symmetries and axiomatic field theory, Brandeis University summer institute in theoretical physics 1965, Gordon and Breach, New York, London, Paris.

[Fe] Fell J.M.G. (1960): *The dual space of C^*-algebras* , Trans. Am. Math. Soc. **94**, 365-403.

[FH] Fredenhagen K. and Hertel J. (1981): *Local algebras of observables and point like localized fields*, Commun. Math. Phys. **80**, 555-561.

[Fo] Fock V. (1932): *Konfigurationsraum und zweite Quantelung*, Z. Physik **75**, 622-653.

[FMS] Fröhlich J., Morchio G. and Strocchi F. (1979): *Infrared problem and spontaneous breaking of the Lorentz group in QED*, Phys.Lett. **89B**, 61-64.

[Go] Godement R. (1947): *Thèoremes taubériens et la theorie spectrale*, Ann. Sci. Ecole. Norm. Sup. **63**, 119-1138.

[Gr] Greenberg O.W. (1961): *Generalized Free Fields and Models of local Field Theory*, Ann. Phys. **16**, 158-176.

[Haa] Haag R. (1959): *Discussion des "axioms" et des propriètés asymptotiques d'une thèorie des champs locale avec partucules composées* , in: les Problèmes mathematiques de la théorie quantique des champs, Pub. CNRS. Vol. **85**.

[HK] Haag R. and Kastler D. (1964): *An algebric approach to quantum field theory*, J. Math. Phys. **5**, 848-861.

[HS] Haag R. and Schroer B. (1962): *Postulates of quantum field theory*, J. Math. Phys. **3**, 248-256.

[HW] Hall D. and Wightman A.S. (1957): *A theorem on invariant analytic function with applications to relativistic quantum field theory*, Mat.-Fys. Medd. Danske Vid. Selsk. **31**, no. 5.

[Heg] Hegerfeldt G.C. (1972): *Gårding domains and analytic vectors for quantum fields* , J. Math. Phys. **13**, 821-827.

[Hei] Heisenberg W. (1925): *Quantentheoretische Umdeutung kinematischer und mechanischer Beziehungen* , Z. f. Phys. **33**, 879-893.

[JW] Jordan P. and Wigner E. (1928): *Über das Paulische Äquivalenzverbot*, Z. f. Physik **47**, 631-651.

[JL] Jost R. and Lehmann H. (1957): *Integraldarstellungen kausaler Kommutatoren* , Nuovo Cimento **5**, 1598-1610.

124

[Kad]   Kadison R.V. (1966): *Derivations of operator algebras* , Ann. of Math. **83**, 280-293.

[Kas]   Kastler D. (1965): *The C^*-algebra of a free boson field I*, Commun. Math. Phys. **1**, 14-48.

[Kö]   Köthe G. (1952): *Die Randwertverteilungen analytischer Funktionen* , Math. Z. **57**, 13-33.

[Kr]   Kraus K. (1970): *An algebraic spectrum condition* , Commun. Math. Phys. **16**, 138-141.

[LR]   Lanford O.E. and Ruelle D. (1967): *Integral Representations of Invariant States on C^*-Algebras* , J. Math. Phys. **8**, 1460-1463.

[Le]   Lehmann H. (1958): *Analytic Properties of Scattering Amplitudes as Functions of Momentum Transfer* , Nuovo Cimento **10**, 579-589.

[Maj]   Majorana E. (1937): *Teoria Simmetrica dell' Elettrone e del Positrone*, Nuovo Cimento **14**, 171.

[Mal]   Malgrange B. (1960): *Division des distributions. I Distributions prolongeables*, in: Seminaire L. Schwartz 1959-60 Exposé **21** Paris.

[MSTV]   Manuceau J., Sirugue M., Testard D. and Verbeure A. (1973): *The smallest C^*-algebra for canonical commutation relations* , Commun. Math. Phys. **32**, 231-243.

[Neu]   Neumann J. von (1931): *Die Eindeutigkeit der Schrödingerschen Operatoren* , Math. Ann. **104**, 570-578.

[Ol]   Olesen D. (1976): *On spectral subspaces and their applications to automorphism groups* , Symposia Math. **20**, 253-572.

[OP]   Olesen D. and Pedersen G.K. (1974): *Derivations of C^*-algebras have simicontinuous generators* , Pacific J. Math. **53**, 563-572.

[Pau]   Pauli W. (1927): *Zur Quantenmechanik des magnetischen Elektrons*, Z. f. Phys. **43**, 601-623.

[Pf74]   Pflug P. (1974): *Über polynomiale Funktionen auf Holomorphiegebieten*, Math. Z. **139**, 133-139.

[Pf82]   Pflug P. (1982): *Eine Bemerkung über die Konstruktion von Holomorphiehüllen* , Zeszyty Naukowe U.J. Prace Matematycne **23**, 21-22.

[Pow]   Powers R.T. (1967): *Representations of uniformly hyperfinite algebras and their associated von Neumann rings* , Ann. Math. **86**, 138-171.

[Roo]   Roos H. (1985): *A Note on Jordan Automorphisms* Ann Inst. Henri Poincaré **42**, 17-29.

[Ru]   Ruelle D. (1966): *States of Physical Systems* , Commun. Math. Phys. **3**, 133-150.

[Sa]   Sakai S. (1966): *Derivations of W^*-algebras* , Ann. of Math. **83**, 273-279.

[SS]   Shale D. and Steinespring W. (1964): *States of the Clifford algebra*, Ann. Math. **80**, 365-381.

[Schw52]   Schwartz L. (1952): *Transformation de Laplace des Distributions* , Comm. Sém Math. Lund, tome suppl. dédié à M. Riesz.

[Schw58]   Schwartz L. (1957/58): *Théorie des distributions valeurs vectorielles* (I), (II) Ann. Inst. Fourier (Grenoble) **7**, 1-141 (1957) **8**, 1-210 (1958).

[Sl]   Slawny J. (1972): *On factor representations and the C^*-algebra of canonical commutation relations* , Commun. Math. Phys. **24**, 151-170.

[Ste]   Stein K. (1937): *Zur Theorie der Funktionen mehrerer komplexer Veränderlicher* , Math. Ann. **114**, 543-569.

[Stø]   Størmer E. (1967): *Large Groups of Automorphisms of C*-Algebras*, Commun. Math. Phys. **5**, 1-22.

[Sto]   Stone M.H. (1932): *On One-Parameter Unitary Groups in Hilbert Space*, Ann. Math. **33**, 643-648.

[Sy]   Symanzik K. (1957): *Derivation of Dispersion Relations for Forward Scattering* Phys. Rev. **105**, 743-749.

[Ti]   Tillmann H.G. (1953): *Randwertverteilungen analytischer Funktionen und Distributionen* , Math. Z. **59**, 61-83.

[Tu]   Turumaru T. (1952): *Crossed product of operator algebras*, Tôhoku Math. J. **4**, 242-251.

[Uhl]   Uhlmann A. (1962): *Über die Definition der Quantenfelder nach Wigtman und Haag*, Wiss. Z. K.M. Univ. Leipzig 11, 213.

[Vl]   Vladimirov V.S. (1960): *The construction of envelopes of holomorphy for domains of special type*, Doklady Akad. Nauk SSSR **134**, 251.

[We]   Weyl H. (1927): *Quantenmechanik und Gruppentheorie* , Z. f. Phys. **46**, 1-46.

[Wi56]   Wightman A.S. (1956): *Quantum field theory in terms of vacuum expectation values*, Phys. Rev. **101**, 860-866.

[Wi60]   Wightman A.S. (1960): *Analytic functions of several complex variables*, in: Relations de dispersion et paticules élémentaires, Ecole d'été Les Houches, C. De Witt and R. Omnes editiors Hermann, Paris.

[Wi64]   Wightman A.S. (1964): *La théorie quantique locale et la théorie quantique des champs*, Ann.Inst.Henri Poincaré A,I, 403-420.

[Y]   Yngvason J. (1989): *Bounded and unbounded realizations of locality*, in: Vth International Congress on Mathematical Physics, Adam Hilger, Bristol and New York.

List of Symbols

Abbreviations: A.=Axiom, C.=Corollary, D.=Definition, L.=Lemma,
N.=Notation, S.=Section, T.=Theorem.

A^n	anti-symmetrization operator on the n-th tensor product, S. I.3
$\mathcal{A}$	C^*-algebra, S. I.1
$\mathcal{A}_m$	short for $\mathcal{A}_{(\mathcal{H}_m, \mathrm{Im}\Delta_m^+)}$, S. I.2
$\mathcal{A}(G)$	C^*-algebra associated with G, S. I.1
$\mathcal{A}(\hat{\mathcal{H}})$	C^*-algebra generated by a Fermi field, S. I.3
$\mathcal{A}_n(\hat{\mathcal{H}})$	n-th grade of the $\mathbb{Z}$-graded algebra $\mathcal{A}(\hat{\mathcal{H}})$, S. I.3
$\{\mathcal{A}, G, \alpha\}$	C^*-dynamical system, S. I.1
$\mathcal{A}(O)$	C^*-algebra associated with O, A. I
$\{\mathcal{A}(O), \mathcal{A}, \mathbb{R}^d, \alpha\}$	theory of local observables, S. I.1
$\{\mathcal{A}(O), \mathcal{A}_m, P, \alpha\}$	local theory of a free field of mass m, L. I.2.2
$\mathcal{A}_{(\mathcal{H}_m, \mathrm{Im}\Delta_m^+)}$	C^*-algebra associated with $\mathcal{H}_m$, S. I.2
$\{\mathcal{A}, \mathbb{R}^d, \alpha\}$	C^*-dynamical system where the group is the translation-group, S. I.1
$\mathcal{A}_{E,\sigma}$	$*$-algebra generated by the symplectic form σ, S. I.2
$\mathcal{A}_{E,\sigma}^1$	completion of a $*$-algebra in an arbitrary norm, S. I.2
$\widehat{\mathcal{A}}_{E,\sigma}$	completion of a $*$-algebra in a C^*-norm, S. I.2
$\mathcal{A}_{E_1, E_2, B}$	$*$-algebra generated by the bilinear form B, S. I.2
$\mathcal{A}_{\mathcal{H},(.,.)}$	$*$-algebra generated by a complex Hilbert space, S. I.2
$\mathcal{A}^{**}$	double dual of $\mathcal{A}$, D. II.2.1
$\mathcal{A}_G^{**}$	the set of elements in $\mathcal{A}^{**}$ which are invariant under the group action, T. II.2.2
$\mathcal{A}^*$	dual–space of $\mathcal{A}$, D. II.2.1
$\mathcal{A}_c^*$	the part of $\mathcal{A}^*$ on which the group acts continuously, D. II.2.1
$\mathcal{A}^*(\mathbb{R}'^+)$	functionals leading to positive energy representations (one-dimensional), D. II.4.3
$\mathcal{A}_0^*(\mathbb{R}'^+)$	functionals with compact energy-support (one-dimensional), D. II.4.3
$\mathcal{A}^*(C')$	functionals leading to positive energy representations (d-dimensional), D. II.6.5
$\mathcal{A}_0^*(C')$	functionals with compact energy-momentum-support (d-dimensional), D. II.6.5
α	a map of the set of observables, S. I.1
α_a	automorphism representing the translation a, A. III
(α, β)	symmetry, S. I.1
α_g	automorphism representing the group element g, A. III.P
α_t	Heisenberg time development, S. I.1
α_φ	automorphism implementing a phase-transformation, S. I.3
B	unit ball $\{z; \|z\| := (\sum |z_i|^2)^{1/2} < 1\}$, S. III.1
Bd	bounded set in $\mathcal{S}(\mathbb{R}^{d-1})$, S. II.1

B_C^+	$= B \cap T(C)$, T. III.2.1
B_C^-	$= B \cap T(-C)$, T. III.2.1
$B(t)$	convex set associated with a tempered distribution, D. II.1.1
$\overset{\circ}{B}(t)$	interior points of $B(t)$, S. II.1
$B(x, y)$	bilinear form, S. I.2
$B = \{b^1, ..., b^d\}$	a basis of $\mathbb{R}^d$, D. II.6.3
B_α	ball of radius r_α centered around x_α, C. III.2.4
B_r	ball of radius r, C. III.2.4
$\mathcal{B}$	algebra of finite sums of products, S. IV.4
	or sub-C^{**}-algebra, S. IV.6
$\{\mathcal{B}(O), \mathcal{B}, \mathbb{R}^d, \alpha\}$	sub–theory of a theory of local observables, T. IV.6.2
β	a map of the sets of states, S. I.1
β	anti-symmetric linear operator, S. I.2
β_t	Schrödinger time development, S. I.1
β_φ	a phase transformation, S. I.3
C	open convex cone, T. II.1.5
C'	closed dual cone with interior points, S. II.1
C_B'	cone with basis B, D. II.6.3
C_x	local cone at the point x, L. III.5.7
$\tilde{C}_x$	set associated with the local cone at x, D. III.5.3
$C^\infty(\mathbb{R}'^d)$	C^∞ functions on $\mathbb{R}'^d$, S. II.3
$C^\infty(x, \mathcal{S}'(x_0))$	C^∞ functions with values in the space of tempered distributions, L. III.4.1
Co	the operator of forming the convex hull, C. III.1.6
$\mathcal{C}(\underline{\mu}, r, t)$	analytic curve depending on certain parameters, S. III.3
$\complement$	complement of a set, S. IV.3
D	unit disc $\{z; |z| < 1\}$, S. III.2
$D^\pm$	unit disc intersected with the upper- lower half-plane respectively, S. III.2
D_r	ball of radius r, L. II.3.4
D_ϵ	ϵ–Neigbourhood of D, L. II.5.2
$D_n(x_0)$	multi-disc centered around x_0, S. III.2
$D_n^\pm(x_0)$	multi-semi-disc centered around x_0, S. III.2
D_n^+	$= D^+ \times D^+ \times ... \times D^+$, C. III.2.3
D_n^-	$= -D_n^+$, C. III.2.3
$D(x_0, ..., x_n)$	the Green function of the wave equation in $n + 1$ dimension, N. III.4.4
$D_{a,b}$	double cone, N. III.3.1
$D_{c,d}^{C_\alpha}$	double cone based on the cone C_α S. III.3
D^i	differential operator of degree i, S. III.4
$d(z)$	distance of z from $\mathbb{C}^n \setminus G$, T. II.1.7
$\mathcal{D}$	space of C^∞-functions with compact support, S. IV.5
$\mathcal{D}y$	set of Dyson functions, T. III.4.5
$\delta(x)$	Dirac's point measure, S. I.2
Δ	subset of the spectrum, S. II.6
$\Delta(\psi)$	a representation for commutator functions, T. III.4.5
$\Delta^+(\psi)$	a representation for the retarded commutator functions, T. III.4.7

$\Delta_m^+(x)$	relativistic Green's function of mass m, S. I.2
$\Delta_m^-(x)$	relativistic Green's function of mass m, S. I.2
$\Delta(z)$	$= \min\{1, d(z), \|z\|^{-1}\}$ generalized distance, S. III.1
$\Delta^\pm$	domains associated with a cone, N. III.3.1
∂	boundary operator, S. II.1
$\partial^\pm$	upper and lower part of the boundary, S. III.6
∂_j	short form of $\frac{\partial}{\partial x_j}$, S. I.3
E	symplectic space, S. I.2
E_0	projection onto all invariant vectors, P. IV.4.5
E_x	subset of the unit sphere, S. III.5
$E(C')$	projection associated with the cone C', D. II.4.2
$E(p)$	projection-valued measure, S. I.1
$E(\lambda)$	right-anihilator of $M((-\infty, -\lambda])$, D. II.4.2
$E(< 0, p >)$	projection associated with an order interval, D. III.6.1
$E(t, \lambda)$	spectral measure of the translation in t-direction, D. II.4.2
E^+	limit of the $E(\lambda)$, D. II.4.1
$E^+(t)$	limit of the $E(t, \lambda)$, D. II.6.1
$\epsilon(I)$	maximum of a difference set, S. II.4
$\epsilon(\lambda)$	$= \mathrm{sign}\lambda$, S. I.3
F	a set in $n + 1$ dimension, D. III.6.2
$F(\pi)$	folium of π-normal states, S. II.7
$F_{x,\psi}(a)$	matrix elements of the translation operator, S. IV.1
$F_{x,\psi}^\pm(a)$	matrix elements of the translation operator, S. IV.1
$F(\Gamma)$	spectral projection, S. II.6
$\tilde{f}$	Fourier transform of f, S. II.3
$\mathcal{F}$	Fourier transformation, S. I.2
$\mathcal{F}_{\mathbb{R}^e}$	partial Fourier transformation, C. II.1.4
$\mathcal{F}$	global field algebra, D. I.3.1
$\mathcal{F}(O)$	field algebra belonging to the domain O, D. I.3.1
$\mathcal{F}_i(O)$	i-th grade of the graded algebra $\mathcal{F}(O)$, L. I.3.2
$\mathcal{F}^n$	subalgebra of the field algebra, N. I.3.3
$\mathcal{F}^n(O)$	subalgebra of the localized field algebra, N. I.3.3
$\{\mathcal{F}^{2n}(O), \mathcal{F}^{2n}, \alpha, \mathbb{R}^4\}$	a model of local algebras based on Fermi-Dirac fields, P. I.3.5
$\Phi(f)$	map from S_r to $\mathcal{H}_m$, S. I.2
φ	linear functional, S. II.1
$|\varphi|$	absolute value of a linear functional, S. II.1
$\varphi(f)$	Fermi–field operator, S. I.3
$(\varphi y)(x)$	$= \varphi(xy)$, S. II.2
$(y\varphi)(x)$	$= \varphi(yx)$, S. II.2
G	arbitrary open domain, S. I.1
G	a group, S. I.1
$\overset{\circ}{G}$	interior of G, S. II.1
$\overline{G}$	closure of G, S. II.1
G'	spacelike complement of a domain, N. III.4.4
$\tilde{G}$	a connected set, D. III.6.2
$\tilde{G}_0$	a set used as first step in a definition, D. III.6.2
$\hat{G}$	extension of a domain into a space of one more dimension, N. III.4.4

$G^{\pm}_{x,\psi}(a)$	matrix elements of the translation operator, S. IV.1
$G(\Gamma, \Delta)$	central carrier of $F(\Gamma)E(\Delta)$, S. II.6
G^{λ}	an interpolating domain, S. III.6
$g_{i,j}$	metric of the Minkowski space, S. I.1
γ^i	Dirac matrices, S. I.3
Γ	a compact set in the dual cone C'_B, S. II.6
Γ	a domain in C^n, S. III.2
$H(G)$	the envelope of holomorphy of G, D. III.1.2
H^+_1	the set $\{z \in \mathbb{C}^n; \mathrm{Im} z_1 > 0\}$, S. III.1
H^+	the upper complex half-plane, S. III.1
$h(u, m)$	hyperboloid with center u and mass m, S. III.4, 71
$\mathcal{H}$	Hilbert space, S. I.1
$\widehat{\mathcal{H}}$	$= \mathcal{H} \oplus \mathcal{H}$, S. I.3
$\mathcal{H}_m$	one–particle Hilbert space in a free field theory, S. I.2
$(\mathcal{H}, \mathrm{Im}(x, y))$	Weyl system, L. I.2.5
$\{\mathcal{H}, \pi, U\}$	covariant representation, S. I.1
$\{\mathcal{H}, \pi, U, \overline{V}^+\}$	representation fulfilling spectrum condition, S. I.1
$\{\mathcal{H}, \pi, U, C'\}$	representation with spectrum in the cone C'
I	index set, S. II.4
I_x	sets for characterization of complex domains, N. III.5.1
K	a cube in $\mathbb{R}^n$, L. III.2.2
K	a conjugation of a complex Hilbert space, S. I.3
$\widehat{K}$	a special conjugation in $\mathcal{H} \oplus \mathcal{H}$, S. I.3
$\mathcal{K}(G)$	set of "commutator functions", T. III.4.5
$\mathcal{L}^2(\mu)$	μ-square integrable functions, S. I.2
$\mathcal{L}(G)$	functions with support in G, S. III.4
(Λ, a)	element of the Poincaré group, S. I.1
m	mass, S. I.2
M	monomial in Dirac operators, D. I.3.1
$\mathcal{M}$	measurement, S. I.1
$M(D)$	set of elements in $\mathcal{A}^{**}$ such that supp $\mathcal{F}^{-1}\varphi(\alpha_a x) \subset D$, D. II.3.2
$\mu(m)$	measure, S. I.2
$\mathcal{M}_c$	von Neumann algebra which is associated to the dual space of $\mathcal{A}^*_c$, S. II.7
$\mathcal{M}_m$	two-sided multiplier of $\mathcal{A}^*_c$, S. II.7
$\mathcal{M}^{(red)}_m$	reduced two-sided multiplier of $\mathcal{A}^*_c$, S. II.7
N	semi–norm on $\mathcal{S}'$, P. II.1.6
N_c	the anihilator of $^*_c$, D. II.3.1
$n(M)$	number of fields in a monomial, D. I.3.1
$n^*(M)$	number of conjugate fields in a monomial, D. I.3.1
$n(x)$	special semi-norm, L. II.3.4
$\mathcal{N}$	complex neighbourhood of a real domain, T. III.2.1
$\nu_{i,j}(p)$	matrix of measures, S. I.2
$\|x\|$	norm of x, S. I.2
O	bounded domain (open), S. I.1
$O(x)$	domain in which an observable x can be measured, S. I.1
$O + a$	translated domain, A. III

O_g	transformed domain, A. III.P
$\mathcal{O}_c$	space of convolution operators on tempered distributions, S. III.4
$\mathcal{O}_m$	space of multiplication operators on tempered distributions, S. III.4
$\mathcal{O}$	set of observables, S. I.1
ω	a state, S. I.1
P	Poincaré group, S. I.1
$P(D)$	a polynomial in the derivatives, S. I.2
$\{P_0, ..., P_{d-1}\}$	generators of the translations, D. IV.2.2
$\mathcal{P}(U)$	set of shifts keeping the spectrum of $U(a)$ in $\overline{V^+}$, D. IV.2.2
π	representation of a C^*-algebra, S. I.1
$\psi(\hat{x})$	a linear map from $\mathcal{H} \oplus \mathcal{H}$ into a C^*-algebra, S. I.3
$\psi(f)$	Dirac field operator, D. I.3.1
$\psi(\hat{f})$	Dirac field operator, D. I.3.1
$< 0, p >$	order interval, D. II.6.1
Q	charge of a sector, S. IV.5
$Q(p)$	polynom in p, S. II.1
$Q^{\pm}(p)$	projections on the mass-hyperboloid, S. I.3
$\{Q_0, ..., Q_{d-1}\}$	generators of translations belonging to the center, D. IV.2.2
$q(\Gamma, \Delta)$	a shift of the spectrum, S. II.6
R	reduced projection for the Dirac equation, S. I.3
$R(D)$	a linear space associated with a domain D, L. II.5.2
$\mathcal{R}$	von Neumann algebra generated by $\pi(\mathcal{A})$ and $U(G)$, P. IV.4.6
spec	spectrum of, S. II.6
supp	support, S. I.2
S	symmetrization operator, S. I.2
S^n	symmetrization operator on the n-particle space, S. I.2
$S_m(x)$	Greens function of the free Dirac field, S. I.3
S_Q	the spectrum of the translations in the sector Q, S. IV.5
$\mathcal{S}$	set of physical states, S. I.1
$\mathcal{S}(\mathbb{R}^d)$	space of strongly decreasing C^∞ functions, S. I.2
$\mathcal{S}_r(\mathbb{R}^d)$	space of real strongly decreasing C^∞ functions, S. I.2
$\mathcal{S}'(\mathbb{R}^d)$	space of tempered distributions, S. II.1
$\mathcal{S}'(\mathbb{R}^e; V)$	space of vector-valued tempered distributions, S. II.1
$\mathcal{S}'(\mathbb{R}^e; \mathcal{S}'(\mathbb{R}^f))$	$= \mathcal{S}'(\mathbb{R}^{e+f})$, S. II.1
$\sigma_x(y)$	a ceiling function, S. III.5
$\sigma(x, y)$	symplectic form, S. I.1
Σ	spacelike surface, N. III.4.4
$\Box$	the wave operator, S. IV.3
$\star$	convolution product, S. II.3
t	distribution, D. III.1.1
$\mathbf{T}$	the one-dimensional torus, S. I.3
$T(V^+)$	forward tube, S. I.2
$T(\overset{\circ}{B}(t))$	tube domain based on $\overset{\circ}{B}(t)$, T. I.1.3
$\Theta(x)$	step function, S. I.2
$\overset{t}{<}$	semi–order in the direction t, D. IV.2.2
$\mathcal{U}$	neigbourhood, S. II.2

U	unitary representation of a group, S. I.1
$U(g)$	unitary representation of the group element g, S. I.1
$U_I(a)$	approximate group representation, S. II.4
$U_B(a)$	a representation of the translations fulfilling the spectrum condition approximately, D. II.6.3
$U(x)$	abbreviation for $\pi(u(x))$, D. I.2.4
$u(x)$	a map of a symplectic space into a $*$-algebra, S. I.2
$\widetilde{u}(x \oplus y)$	a map fom $E \oplus E$ into a $*$-algebra, S. I.2
$\mathcal{V}$	vector space, S. II.1
V^+	open forward light–cone, S. I.1
$\widehat{V}^+$	forward light–cone in a space of one more dimension, N. III.4.4
$\overline{V}^+$	closed forward light–cone, S. I.1
$V(g)$	unitary group representation, P I.2.6
$v(x)$	a map of a linear space into a $*$-algebra, S. I.2
$W(a)$	group representation belonging, to the center, S. II.6
$x(f)$	smeared operator for elements belonging to $x \in \mathcal{A}^{**}E^+$, S. II.4
$[x(f)]$	equivalence class of operators, D. II.3.1
$x(\underline{\mu}, r, t)$	parametrization of a hyperboloid, S. III.3
(x, y)	scalar product in a complex Hilbert space, S. I.2
$< x, y >$	scalar product in a real Hilbert space, S. I.2
$\mathcal{Z}(\mathcal{A})$	center of $\mathcal{A}$, P. III.4.2

Springer-Verlag
and the Environment

We at Springer-Verlag firmly believe that an international science publisher has a special obligation to the environment, and our corporate policies consistently reflect this conviction.

We also expect our business partners – paper mills, printers, packaging manufacturers, etc. – to commit themselves to using environmentally friendly materials and production processes.

The paper in this book is made from low- or no-chlorine pulp and is acid free, in conformance with international standards for paper permanency.

Lecture Notes in Physics

For information about Vols. 1–434
please contact your bookseller or Springer-Verlag

Vol. 435: E. Maruyama, H. Watanabe (Eds.), Physics and Industry. Proceedings, 1993. VII, 108 pages. 1994.

Vol. 436: A. Alekseev, A. Hietamäki, K. Huitu, A. Morozov, A. Niemi (Eds.), Integrable Models and Strings. Proceedings, 1993. VII, 280 pages. 1994.

Vol. 437: K. K. Bardhan, B. K. Chakrabarti, A. Hansen (Eds.), Non-Linearity and Breakdown in Soft Condensed Matter. Proceedings, 1993. XI, 340 pages. 1994.

Vol. 438: A. Pękalski (Ed.), Diffusion Processes: Experiment, Theory, Simulations. Proceedings, 1994. VIII, 312 pages. 1994.

Vol. 439: T. L. Wilson, K. J. Johnston (Eds.), The Structure and Content of Molecular Clouds. 25 Years of Molecular Radioastronomy. Proceedings, 1993. XIII, 308 pages. 1994.

Vol. 440: H. Latal, W. Schweiger (Eds.), Matter Under Extreme Conditions. Proceedings, 1994. IX, 243 pages. 1994.

Vol. 441: J. M. Arias, M. I. Gallardo, M. Lozano (Eds.), Response of the Nuclear System to External Forces. Proceedings, 1994, VIII. 293 pages. 1995.

Vol. 442: P. A. Bois, E. Dériat, R. Gatignol, A. Rigolot (Eds.), Asymptotic Modelling in Fluid Mechanics. Proceedings, 1994. XII, 307 pages. 1995.

Vol. 443: D. Koester, K. Werner (Eds.), White Dwarfs. Proceedings, 1994. XII, 348 pages. 1995.

Vol. 444: A. O. Benz, A. Krüger (Eds.), Coronal Magnetic Energy Releases. Proceedings, 1994. X, 293 pages. 1995.

Vol. 445: J. Brey, J. Marro, J. M. Rubí, M. San Miguel (Eds.), 25 Years of Non-Equilibrium Statistical Mechanics. Proceedings, 1994. XVII, 387 pages. 1995.

Vol. 446: V. Rivasseau (Ed.), Constructive Physics. Results in Field Theory, Statistical Mechanics and Condensed Matter Physics. Proceedings, 1994. X, 337 pages. 1995.

Vol. 447: G. Aktaş, C. Saçlıoğlu, M. Serdaroğlu (Eds.), Strings and Symmetries. Proceedings, 1994. XIV, 389 pages. 1995.

Vol. 448: P. L. Garrido, J. Marro (Eds.), Third Granada Lectures in Computational Physics. Proceedings, 1994. XIV, 346 pages. 1995.

Vol. 449: J. Buckmaster, T. Takeno (Eds.), Modeling in Combustion Science. Proceedings, 1994. X, 369 pages. 1995.

Vol. 450: M. F. Shlesinger, G. M. Zaslavsky, U. Frisch (Eds.), Lévy Flights and Related Topics in Physics. Proceedings, 1994. XIV, 347 pages. 1995.

Vol. 451: P. Krée, W. Wedig (Eds.), Probabilistic Methods in Applied Physics. IX, 393 pages. 1995.

Vol. 452: A. M. Bernstein, B. R. Holstein (Eds.), Chiral Dynamics: Theory and Experiment. Proceedings, 1994. VIII, 351 pages. 1995.

Vol. 453: S. M. Deshpande, S. S. Desai, R. Narasimha (Eds.), Fourteenth International Conference on Numerical Methods in Fluid Dynamics. Proceedings, 1994. XIII, 589 pages. 1995.

Vol. 454: J. Greiner, H. W. Duerbeck, R. E. Gershberg (Eds.), Flares and Flashes, Germany 1994. XXII, 477 pages. 1995.

Vol. 455: F. Occhionero (Ed.), Birth of the Universe and Fundamental Physics. Proceedings, 1994. XV, 387 pages. 1995.

Vol. 456: H. B. Geyer (Ed.), Field Theory, Topology and Condensed Matter Physics. Proceedings, 1994. XII, 206 pages. 1995.

Vol. 457: P. Garbaczewski, M. Wolf, A. Weron (Eds.), Chaos – The Interplay Between Stochastic and Deterministic Behaviour. Proceedings, 1995. XII, 573 pages. 1995.

Vol. 458: I. W. Roxburgh, J.-L. Masnou (Eds.), Physical Processes in Astrophysics. Proceedings, 1993. XII, 249 pages. 1995.

Vol. 459: G. Winnewisser, G. C. Pelz (Eds.), The Physics and Chemistry of Interstellar Molecular Clouds. Proceedings, 1993. XV, 393 pages. 1995.

Vol. 460: S. Cotsakis, G. W. Gibbons (Eds.), Global Structure and Evolution in General Relativity. Proceedings, 1994. IX, 173 pages. 1996.

Vol. 461: R. López-Peña, R. Capovilla, R. García-Pelayo, H. Waelbroeck, F. Zertuche (Eds.), Complex Systems and Binary Networks. Lectures, México 1995. X, 223 pages. 1995.

Vol. 462: M. Meneguzzi, A. Pouquet, P.-L. Sulem (Eds.), Small-Scale Structures in Three-Dimensional Hydrodynamic and Magnetohydrodynamic Turbulence. Proceedings, 1995. IX, 421 pages. 1995.

Vol. 463: H. Hippelein, K. Meisenheimer, H.-J. Röser (Eds.), Galaxies in the Young Universe. Proceedings, 1994. XV, 314 pages. 1995.

Vol. 464: L. Ratke, H. U. Walter, B. Feuerbach (Eds.), Materials and Fluids Under Low Gravity. Proceedings, 1994. XVIII, 424 pages, 1996.

Vol. 466: H. Ebert, G. Schütz (Eds.), Spin – Orbit-Influenced Spectroscopies of Magnetic Solids. Proceedings, 1995. VII, 287 pages, 1996.

Vol. 467: A. Steinchen (Ed.), Dynamics of Multiphase Flows Across Interfaces. XII, 267 pages. 1996.

Vol. 468: C. Chiuderi, G. Einaudi (Eds.), Plasma Astrophysics. 1994. VII, 326 pages. 1996.

New Series m: Monographs

Vol. m 1: H. Hora, Plasmas at High Temperature and Density. VIII, 442 pages. 1991.

Vol. m 2: P. Busch, P. J. Lahti, P. Mittelstaedt, The Quantum Theory of Measurement. XIII, 165 pages. 1991.

Vol. m 3: A. Heck, J. M. Perdang (Eds.), Applying Fractals in Astronomy. IX, 210 pages. 1991.

Vol. m 4: R. K. Zeytounian, Mécanique des fluides fondamentale. XV, 615 pages, 1991.

Vol. m 5: R. K. Zeytounian, Meteorological Fluid Dynamics. XI, 346 pages. 1991.

Vol. m 6: N. M. J. Woodhouse, Special Relativity. VIII, 86 pages. 1992.

Vol. m 7: G. Morandi, The Role of Topology in Classical and Quantum Physics. XIII, 239 pages. 1992.

Vol. m 8: D. Funaro, Polynomial Approximation of Differential Equations. X, 305 pages. 1992.

Vol. m 9: M. Namiki, Stochastic Quantization. X, 217 pages. 1992.

Vol. m 10: J. Hoppe, Lectures on Integrable Systems. VII, 111 pages. 1992.

Vol. m 11: A. D. Yaghjian, Relativistic Dynamics of a Charged Sphere. XII, 115 pages. 1992.

Vol. m 12: G. Esposito, Quantum Gravity, Quantum Cosmology and Lorentzian Geometries. Second Corrected and Enlarged Edition. XVIII, 349 pages. 1994.

Vol. m 13: M. Klein, A. Knauf, Classical Planar Scattering by Coulombic Potentials. V, 142 pages. 1992.

Vol. m 14: A. Lerda, Anyons. XI, 138 pages. 1992.

Vol. m 15: N. Peters, B. Rogg (Eds.), Reduced Kinetic Mechanisms for Applications in Combustion Systems. X, 360 pages. 1993.

Vol. m 16: P. Christe, M. Henkel, Introduction to Conformal Invariance and Its Applications to Critical Phenomena. XV, 260 pages. 1993.

Vol. m 17: M. Schoen, Computer Simulation of Condensed Phases in Complex Geometries. X, 136 pages. 1993.

Vol. m 18: H. Carmichael, An Open Systems Approach to Quantum Optics. X, 179 pages. 1993.

Vol. m 19: S. D. Bogan, M. K. Hinders, Interface Effects in Elastic Wave Scattering. XII, 182 pages. 1994.

Vol. m 20: E. Abdalla, M. C. B. Abdalla, D. Dalmazi, A. Zadra, 2D-Gravity in Non-Critical Strings. IX, 319 pages. 1994.

Vol. m 21: G. P. Berman, E. N. Bulgakov, D. D. Holm, Crossover-Time in Quantum Boson and Spin Systems. XI, 268 pages. 1994.

Vol. m 22: M.-O. Hongler, Chaotic and Stochastic Behaviour in Automatic Production Lines. V, 85 pages. 1994.

Vol. m 23: V. S. Viswanath, G. Müller, The Recursion Method. X, 259 pages. 1994.

Vol. m 24: A. Ern, V. Giovangigli, Multicomponent Transport Algorithms. XIV, 427 pages. 1994.

Vol. m 25: A. V. Bogdanov, G. V. Dubrovskiy, M. P. Krutikov, D. V. Kulginov, V. M. Strelchenya, Interaction of Gases with Surfaces. XIV, 132 pages. 1995.

Vol. m 26: M. Dineykhan, G. V. Efimov, G. Ganbold, S. N. Nedelko, Oscillator Representation in Quantum Physics. IX, 279 pages. 1995.

Vol. m 27: J. T. Ottesen, Infinite Dimensional Groups and Algebras in Quantum Physics. IX, 218 pages. 1995.

Vol. m 28: O. Piguet, S. P. Sorella, Algebraic Renormalization. IX, 134 pages. 1995.

Vol. m 29: C. Bendjaballah, Introduction to Photon Communication. VII, 193 pages. 1995.

Vol. m 30: A. J. Greer, W. J. Kossler, Low Magnetic Fields in Anisotropic Superconductors. VII, 161 pages. 1995.

Vol. m 31: P. Busch, M. Grabowski, P. J. Lahti, Operational Quantum Physics. XI, 230 pages. 1995.

Vol. m 32: L. de Broglie, Diverses questions de mécanique et de thermodynamique classiques et relativistes. XII, 198 pages. 1995.

Vol. m 33: R. Alkofer, H. Reinhardt, Chiral Quark Dynamics. VIII, 115 pages. 1995.

Vol. m 34: R. Jost, Das Märchen vom Elfenbeinernen Turm. VIII, 286 pages. 1995.

Vol. m 35: E. Elizalde, Ten Physical Applications of Spectral Zeta Functions. XIV, 228 pages. 1995.

Vol. m 36: G. Dunne, Self-Dual Chern-Simons Theories. X, 217 pages. 1995.

Vol. m 37: S. Childress, A.D. Gilbert, Stretch, Twist, Fold: The Fast Dynamo. XI, 410 pages. 1995.

Vol. m 38: J. González, M. A. Martín-Delgado, G. Sierra, A. H. Vozmediano, Quantum Electron Liquids and High-T_c Superconductivity. X, 299 pages. 1995.

Vol. m 39: L. Pittner, Algebraic Foundations of Non-Commutative Differential Geometry and Quantum Groups. XII, 469 pages. 1996.

Vol. m 40: H.-J. Borchers, Translation Group and Particle Representations in Quantum Field Theory. VII, 131 pages. 1996.